Improving Regulatory Delivery in Food Safety

MITIGATING OLD AND NEW RISKS, AND FOSTERING RECOVERY

OECD

BETTER POLICIES FOR BETTER LIVES

This document, as well as any data and map included herein, are without prejudice to the status of or sovereignty over any territory, to the delimitation of international frontiers and boundaries and to the name of any territory, city or area.

Please cite this publication as:
OECD (2021), *Improving Regulatory Delivery in Food Safety: Mitigating Old and New Risks, and Fostering Recovery*, OECD Publishing, Paris, *https://doi.org/10.1787/bf34907e-en*.

ISBN 978-92-64-45605-1 (print)
ISBN 978-92-64-55424-5 (pdf)

Foreword

This report presents a collection of regulatory responses at regional, national and international levels to ensure food safety during the pandemic. The pandemic brought additional challenges in the food safety domain in both advanced and less advanced economies. Advanced economies, where food security is not usually a challenge, had not felt the scale of such a sanitary emergency in recent history – this made managing the crisis all the more challenging. In less advanced economies, on the other hand, infrastructure and technological limitations have made economic recovery more arduous. Despite the resilience displayed by food supply chains, the pandemic has placed several additional burdens on food business operators to ensure that food supplies continue at pre-pandemic levels while applying additional lockdown, social-distancing and safety rules. Emergency laws were passed in several jurisdictions, including the EU, to prevent regulatory processes from being unduly hampered or from unjustifiably hindering business activities.

In addition to the evaluation of fast-tracked regulations passed during the pandemic and to the assessment of burdens, regulators willing to secure economic recovery will have to strike a delicate balance between reducing administrative burdens and ensuring high levels of food safety. As has been found in this study, reducing inspections, introducing self-compliance models or making inspections remote have not reduced food safety. Yet, moving forward, regulators and policy makers will have to ascertain how to apply and enforce rules in a post-pandemic world. Questions related to food safety are often complex and, at times, strategically important. This is because the risks associated with food safety can be multidimensional, linked to other socio-economic and environmental factors. This is perhaps one of the important lessons from the COVID-19 pandemic.

Whatever the scenario, the COVID crisis has once again stressed the need for co-operation and risk-based, simpler and proportionate regulation. A risk-based assessment of current control plans and activities will need to be carried out to ascertain whether they are fit for purpose and can withstand the fallout from a global crisis.

This report was approved by the OECD Regulatory Policy Committee through written procedure on 8 June 2021 and prepared for publication by the OECD Secretariat.

Acknowledgements

This report was produced with grant support from the Global Food Safety Initiative (GFSI) of the Consumer Goods Forum (CGF).

This report was prepared by the Directorate for Public Governance (GOV), under the leadership of Elsa Pilichowski, Director, and Nick Malyshev, Head of the Regulatory Policy Division in GOV.

This report was authored by Florentin Blanc, Senior Policy Analyst, along with Giuseppa Ottimofiore and Camila Saffirio, Policy Analysts, of the Regulatory Policy Division. The first chapter on "food safety challenges, informal markets, and their role in the crisis" was authored by Donald Macrae (consultant). The authors would like to thank Francesco Calisi, Carola Bertone and Hamsini Shankar, consultants, for their support. The report was prepared for publication by Jennifer Stein.

The authors are also thankful to colleagues from within the OECD for their feedback and comments, including Elena Avery and team from the Trade and Agriculture Directorate, Soria Morales Ernesto and Carina Lindberg team from Policy Coherence for Sustainable Development (GOV). The authors would also like to thank Peter Wend from Bundesamt für Verbraucherschutz und Lebensmittelsicherheit and Filippo Castoldi from Servizi Veterinari della Regione Lombardia, Italy for their valuable inputs.

Table of contents

Tables

Figures

Boxes

Abbreviations and acronyms

CAP	Conformity assessment procedure
CFIA	Canadian Food Inspection Agency
CGMPs	Current good manufacturing practices
CDC	Centers for Disease Control and Prevention, United States
FDA	Food and Drug Administration, United States
FSA	Food Standards Agency, United Kingdom
FSSAI	Food Safety and Standards Authority for India
GFSI	Global Food Safety Initiative
HACCP	Hazard analysis and critical control points
HSE	Health and Safety Executive, United Kingdom
IAF	International Accreditation Forum
ICT	Information Communication Technology
ILAC	International Laboratory Accreditation Cooperation
ISO	International Standardisation Organization
NQI	National Quality Infrastructure
OECD	Organisation for Economic Co-operation and Development
OIE	World Organisation for Animal Health
OSHA	Occupational Safety and Health Administration, United States
PCSD	Policy Coherence for Sustainable Development
RASFF	Rapid Alert System for Food and Feed, European Union
RIA	Regulatory Impact Assessment
SDGs	Sustainable Development Goals
SPS	Sanitary and Phytosanitary
TBT	Technical Barriers to Trade
TRACES	Trade Control and Expert System, European Union
WHO	World Health Organization

Executive summary

The COVID-19 pandemic has revealed vulnerabilities in safety in both the food chain and food supplies

The pandemic highlighted how developing nations play a crucial role in ensuring food chain safety and how rules and sanctions alone cannot achieve desired safety goals. Food plays an intricate role amongst human beings, and also reflects a society's socio-economic and environmental condition. A wet market in China, where the novel corona virus is thought to have originated, raises the interesting debate of "supermarketisation" – a phenomenon where the private sector (super/hyper markets) contributes significantly to strengthening food value chains – and the rise of the elite. "Supermarketisation" is a situation where, influenced by the process of urbanization and income growth, traditional retail outlets and wet markets are increasingly replaced by modern retail outlets such as supermarkets. While wet markets are known to be a constant threat of zoonotic diseases, simply bringing in regulations to curb the spread of diseases from wet markets may not be effective. This is because wet markets are more than just a source of food for millions of poor people. They are invariably informal in their organisation and fundamental to the socio-economic functioning of the local communities within which they operate. Supermarkets are thought to disrupt this harmony. Therefore, first, the design of regulations needs to factor in these local contexts, for a food safety regulation may have unintended impacts in other areas. Second, achieving food safety requires a robust implementation infrastructure. Food safety regulatory delivery should rely as much as possible on "enabling and engaging" action that helps businesses focus on the "good" rather than just prevent the "bad".

Changing behaviour through cultural shifts can go a long way towards achieving and ensuring food safety

Organisational behaviour, especially among managers and workers, has taken on greater importance in the debate on making regulations more responsive to human behaviour. Despite several regulatory controls in place, the threat of foodborne diseases always exists. This could be because businesses often forego safety precautions for profit or sometimes simply because employees underestimate the risks or do not have a clear understanding of the rules. To combat this, creating a culture of hygiene across businesses can go a long way in controlling the spread of infectious diseases. Top-down approaches in corporate structures hinder creativity and, in turn, reduce initiatives by workers. It is important to increase workers' understanding of why they do what they do. This can ensure a better application of rules from theory to practice. Success stories show that companies that have a stronger food safety culture do a better job at implementing new standards or responding to new threats. Regulators should, however, be mindful of the characteristics of the business in question. Safety culture requirements must take into account the nature and size of the business. A prior assessment of the prevailing culture and the culture maturity matrix can also help regulators frame better policies. The benefits of a strong food safety culture are known; the challenge lies in implementing food safety cultures using evaluation tools for control agencies as well as businesses.

Digital tools and flexibility can ensure seamless regulatory delivery in times of crisis

The COVID crisis required governments across the world to take swift action and adjust their regulatory delivery – or the way their regulatory agencies operate to achieve the intended outcomes of regulation. This adjustment had to be balanced with the need to ensure the smooth functioning of the food supply chain. The pandemic, briefly, also saw changes in consumer trends. For instance, more food was supplied to supermarkets because restaurants were closed. Consumers also shifted to online shopping as a convenient and safer way of obtaining groceries. This added additional burdens on authorities who now had to monitor the sale of food supplies online. Difficulties in carrying out physical inspections, a lack of protective equipment, the reshuffling of inspectorates to respond to the pandemic all resulted in an easing of regulatory measures through temporary legislation. This was done in multiple ways – deferred inspections, e-certifications, lenient approaches to labelling, third-party involvement in audit and certification processes, etc. Regulators across the world see an increased potential in using digital tools such as machine learning and big data to improve regulatory capacity. Using these tools, inspectors can sift through volumes of data and perform risk-based assessments. While these tools were certainly useful during the pandemic, the future scope and use of them raise questions. For example, it has been found that remote inspections are not always possible. In some cases, remote inspections and audits were not found to be helpful, for instance for the initial assessment of food hygiene in a new operator, or in high-risk food businesses (where there is a higher public health risk associated with failing to identify food hazards) and it was found more appropriate to continue relying on "on site" controls instead. In order to move forward and to foster economic recovery, regulators will have to consider regulatory impact assessments of temporary measures, hybrid approaches, greater risk-based prioritisation and, most importantly, improving co-ordination in inspection functions.

Economic recovery will need optimising of regulatory systems

While the pandemic has created its own set of challenges and put into question existing regulatory delivery performance, it has also opened new opportunities for regulators to rethink and reprioritise their approach to regulatory delivery. One way forward is to have a risk-based and flexible regulatory system. Risk-based regulatory systems have been performing risk analysis of materials and processes that need to be changed and updating their systems to accommodate the results of their risk analysis. Regulators must be flexible and agile to come up with temporary solutions, whether risk-based or not. Inefficiencies related to fragmentation and duplication of regulatory responses also have to be addressed. Finally, lessons learnt in times of crisis should be applied in normal times. Planning and prioritisation of controls is important, and mock drills during normal times can prove useful in assessing human as well as infrastructure capacities.

Introduction

The reliability and safety of the food supply are fundamental to the existence and preservation of human societies, and food safety thus remains a prominent concern of citizens and policy makers, even in advanced economies where the quantity of supply is of less concern. Thus, from the onset of the COVID-19 crisis, ensuring that food supplies remain both reliable and safe has been a key issue, even if made considerably less visible by the urgency of the health crisis created by the virus. Regulatory systems have had a critical role to play in this context – ensuring continued safety, while removing (as much as possible) barriers that may make it more difficult for the industry to continue operating in spite of the crisis.

The origin of the pandemic is (though still largely uncertain) probably connected to the food chain, with virus transmission from animals to humans apparently linked to a food market in Wuhan. The presence of live animals, and of species presenting particular health risks, is in itself an issue that corresponds to difficulties in improving food safety (and implementation of food safety rules), in China and other emerging markets. Moreover, the consumption of "wild" meat may itself be linked to shortages in the supply of other meats, possibly linked to challenges in animal health or to the fact that wild meat is often an affordable primary source of animal protein and a source of cash revenue to economically weaker sections in several developing nations. Reliance on wild meat may also have been triggered by ecological reasons such as deforestation and loss of food resources. Again, even if the precise causal links for COVID-19 emergence are not fully ascertained, there have been a number of other zoonotic episodes in China and other developing countries, showing the importance of the food chain for epidemiological risks.

Managing the crisis and limiting pandemic spread has proved very difficult, particularly in countries that did not have recent experience with major epidemics of respiratory diseases. Changing behaviour at work and in social life has been challenging, leading to repeated spikes in contaminations and fatalities. Learning from existing experience in supporting behaviour change is thus of particular importance, and the food industry (with major change in practices over three decades) is a valuable source of such learning.

Despite the resilience the food supply chain showed during the pandemic, the events of the past year have also placed a great burden on food business operators (and specifically their workers) to continue supplying the population in spite of lockdowns and quarantines, and/or while complying with new rules on physical distancing and contagion prevention that may require significant changes in production processes. In some cases, major outbreaks have taken place at food production facilities, showing the difficulty to reconcile occupational health / public health with food supply requirements in this crisis situation. It also has created new challenges for state regulators and private certifiers alike, in terms of how to operate in times of lockdown and social distancing.

Human health and resilience need to be strengthened against lifestyle diseases such as heart diseases and cancers (which are linked to the food we eat) and reduce immune response to illnesses such as COVID-19. In addition to this, economic recovery will also have to be promoted by supporting in all possible ways the food sector, including through reduction of administrative burden – while ensuring food safety in the most efficient and effective way possible through a risk-based, compliance-supporting regulatory delivery system.

This report looks, successively, at the links between food safety and animal health, and pandemic emergence, and what we can learn from this to try and improve food safety in a developing/transition context and contribute to managing better zoonotic risks. It then considers how the lessons learnt from spreading a "culture of food safety" over the past couple of decades helped in fostering behaviour change to better reduce pandemic risks, at a time when the COVID-19 crisis has shown how difficult it can be to achieve widespread safer conduct throughout industries and society. In its third chapter, it considers how regulatory systems as well as third-party certification systems have adapted and responded to the crisis. The final chapter looks into how regulatory reform in the food safety sphere can support economic recovery and resilience in a sector, which is particularly crucial from a social perspective.

1 Food safety challenges, informal markets, and their role in the crisis

In this opening chapter, we look at the interaction between the COVID-19 crisis and the food chain. While the pandemic is not a "food safety issue" in the strict sense, its emergence is linked in several ways to the food chain – and to safety of supplies, in particular veterinary issues. Regardless of the still imperfectly known mechanism of SARS-Cov-2 spread to humans, understanding better the ways in which zoonoses can lead to human pandemics, as well as how regulatory systems can help with reducing such risks, is essential. Particularly essential is to identify approaches through which regulatory systems in developing and emerging economies can support improved safety in the food chain, in contexts where imposing costly rules and rigid mechanisms are likely to be of limited effectiveness.

Introduction

The pandemic level spread of the COVID-19 virus has generated concern and confusion, as people all over the world have tried to make sense of what was (and still is) happening, with some looking to attach blame to human agents as proxies for the invisible and totally unconcerned virus itself. A common version of the origin of the problem is that a human ate a wild animal bought in a wet market in China and acted as a conduit for this new virus to jump species from wild animals to humans. This characterises the issue as a failure of food safety regulation for allowing that danger into the food chain. This chapter examines the hypotheses about the origin of the virus and then explores the challenges to a food safety regulatory system in coping with the threat of a pandemic such as this.

First, it sets out the current scientific understanding of the origin of the virus, that it is a new version from an existing family of viruses which has mutated within a genetic reservoir in the wild, probably amongst bats, and has transferred from animals to humans. It is particularly well adapted to humans and spreads easily between humans, leading to the pandemic level of infection. The exact point of the "spill over" from animals to humans and the method of transmission is not known and is subject to some controversy. A human eating a wild animal that was infected by infected bats is a possibility, but the virus could also have transferred by touch or inhalation, without having anything to do with food or eating. The spill over may have been in the wet market in Wuhan or the spread of the virus may have been amplified by human interactions in that market but there is evidence of early cases of the infection in humans connected with that market *before* the "outbreak".

It took fifteen years to establish the origin of an earlier virus, so it is unwise to try and be definitive at this stage about what happened as to the origin of this virus. The World Health Organisation (WHO) of the United Nations has set up an investigation, but there is much still left to investigate, and there is debate as to whether sufficient access is currently available for investigative work to be fully effective (Zarocostas, 2021[1]). Scientific research does attest to circulation of coronaviruses related to the SARS-Cov-2 of the Covid-19 pandemic in wild animals in Southern China and nearby countries (Wacharapluesadee et al., 2021[2]), but so far the precise details of the pandemic origins are still highly uncertain. But lack of the actual detail of what happened does not invalidate discussion about the relationship between food safety and pandemics such as this. Although evidence of actual connection at source is not clear, the course of this pandemic – coming after a few others – increases concerns about some parts of the food chain, in terms of security of supply and safety of the production and supply processes. These concerns should inform actions that may reduce some of the hazards that provide a context for the growth of another pandemic.

Food is so fundamental to human existence that it connects with many different areas of activity. The impact on food of diseases in animals is just one. Transmission of diseases from animals can be through food and also through other direct and indirect pathways. The possibility of one route from pathogen to food contamination does not then mean that what may be necessary to prevent that possibility should be exclusively or primarily a matter of food safety and specifically of food safety *regulation*. That said, food safety can often have bidirectional links with other development goals, in that food safety is seen as having a direct role as well as an enabling role for several of the Sustainable Development Goals (SDGs). An example of this is where food safety can directly enable SDG 3 on good health and well-being. The reverse is also true like when SDGs not directly related to food safety- like SDGs 7, 13 and 14[1] can, in fact, have a strengthening effect on food safety. One could argue that tackling marine pollution and promoting climate action has a bearing on food derived from marine sources and in turn can have an enabling effect on food safety. One could also assert that SDG 4 on quality education could strengthen a case for wholistic and safe nutrition from animal sources and required for cognitive development in children. Yet, on the other hand, food safety regulations can also add unnecessary burdens on achieving SDGs. For instance, food safety regulations, as non-tariff measures, can harm food exporters from lesser developed nations and thereby reduce income generation. The synergies and trade-offs are important to understand for greater policy coherence. Food safety is an integral part of a broader and complex strategy of managing other challenges and the context in which a risk related to food safety arises may have significant socio-economic and environmental elements or may require a different way of organising a rural economy or a method of agriculture.

The paper focuses on the role of wet markets and the debate in the developing world around what is referred to as "supermarketisation"[2] as an illustration of how easily a food safety question strays into very complex and strategic issues. The OECD (OECD, 2021[3]) has highlighted the triple challenge food systems around the world have faced, even before the pandemic. All these three issues are complex and involve i) food security and nutrition ii) livelihoods and nutrition and iii) resource use and climate change. Policies directed at resolving these challenges come with their own synergies and trade-offs, and policy makers need to be cognizant of this. Reorganisation of rural economies, trade-offs related to farm income vs consumer prices or lowering of livestock (and consequentially protein supply) to control emissions, does not mean that these issues become part of the remit of food safety regulation, but that food safety can be *one factor* in various complex issues, whether as a consequence or as a cause of that complexity. That can sometimes lead to regulatory changes within the food safety remit which can then *contribute* to solving the wider problem (but can be no more than a *part* of the solution).

Boundaries have to be set on the scope of what is regarded as "food safety". This is particularly the case when food safety is seen as a regulatory framework, as opposed to a branch of scientific research or a theory of development economics. As a matter of law, the scope of the regulations has to be limited and clear. But a regulatory system is more than just an abstract framework of laws and regulations. It may also have an institutional infrastructure for implementation, with significant resources and techniques for delivering change in the real world. The scope of that delivery infrastructure is arguably the key issue. Perhaps more than any other regulatory system, food safety has been wrestling with the challenge of changing common habits across huge populations. Delivering the intended regulatory outcomes of food safety regulation has become more than just delivering compliance with technical rules, especially in countries where the informal, unregulated markets provide the majority of the food consumed – and where "traditional" production methods and supply chains may predominate.

This chapter takes one such delivery infrastructure as an example – a very specific one, but highly relevant to the context of the pandemic origins: the food safety authority for India, facing the challenge of ensuring safe and wholesome food for a significant percentage of the entire world population (see Delivering food safety). It illustrates how the delivery of regulatory objectives has been transformed, from "enact and enforce" to "engage and enable". The actual regulations have been given a different perspective to the traditional assumption of a top-down instruction that will be obeyed and will solve the problem. Their role remains as a floor for implementation policy and as an indicator of what needs to be achieved but implementation goes much wider, especially for food safety. Much of the world's food comes from informal markets that are, by definition, unregulated but food safety regulatory delivery organisations can still apply techniques to raise practice standards. However, a key element in this modern approach is that these delivery organisations can partner with others in multidisciplinary networks to help solve some of the intractable and complex issues involving safe food, without having to restructure government Ministries and subsidiary institutions. The fundamental question is "how to improve safety", and formal rules and regulatory systems are just one element of the answer (Blanc and Macrae, 2021[4]).

The emerging picture of the pandemic and its connections with the food chain

There is nothing new or unusual about humans becoming ill from pathogens hosted by animals. A Zoonotic Diseases Fact Sheet[3] lists 26 diseases that can transmit to humans from animals (called "zoonoses" or "zoonotic diseases"). They include lethal diseases like rabies or malaria and also pathogens such as salmonella, brucellosis and shigella. Humans and animals have been co-existing for millennia and also eating each other, so the spread of disease from one to another has been common. However, it has also been increasingly predictable and avoidable, precisely because it has been so common. Treatments have been developed to reduce the impact and practices have been applied to reduce transmission. But occasionally new zoonotic pathogens appear which have no pre-existing medical treatments and may also be lethal. If transmission can then be from human to human, once it has crossed the species barrier, it can become a pandemic.

The original transmission to a human is important in understanding the virus but of limited importance in managing its spread. This is because pathogenic organisms mutate over time and often become more virulent and harder to control. It is a coronavirus, a family of viruses that have genetic origins up to 10 000 years ago, although a common ancestor for all coronaviruses could go back millions of years (Wertheim et al., 2013[5]). Bats are seen as a natural reservoir for coronaviruses but recent transmission of novel coronaviruses to humans may have been through an intermediate host since there is little close contact between humans and bats (World Health Organization, 2020[6]). What is "novel" about this virus is that it is a new mutation of the coronavirus family, which is likely to have occurred within the genetic reservoir of host animals. SARS-CoV-2, the novel virus responsible for COVID-19, is not found in farmed or domestic animals, although they could act as an intermediary between bats and humans. The intermediary could also be a wild animal.

The source of the COVID-19 transmission is uncertain and contentious. Cases of pneumonia-type illness were identified in Wuhan, China, in December 2019 and it was identified as what is now called COVID-19. There was a clear connection between these cases and the Huanan Wholesale Seafood Market in Wuhan City and so this was assumed by many to be the source. The market sold both farmed and wild animals. This combination of factors led to an assumption that the original "spill over" from animals to humans arose from wild animals sold in that market. However, this has been challenged and the WHO is investigating the origin of the virus (World Health Organization, 2020[7]). There have been reports of cases of the virus being identified in Europe prior to the cases in Wuhan. One claims that a test of 959 people in Italy in a lung cancer screening trial between September 2019 and March 2020 showed antibodies to the virus present in some September results, although it was also argued that this did not prove that the virus did not originate in China but only that it did not start in December with the cases in Wuhan (Parodi E, 2020[8]). A report by the BBC refers to a study of wastewater from Milan and Turin on 18 December 2019 which had traces of the virus and also refers to indicators in France and Spain that the virus was present before cases in these countries were confirmed (BBC News, 2020[9]). The WHO investigation into the origin acknowledges that it can be very difficult to trace back a zoonotic spill over, recognising that it could take years.

It is possible that the COVID-19 pandemic, that has had such globally damaging results originated in humans eating wild animals which carried this mutation of the coronavirus family. That would make this the most serious outbreak of food-based illness in history. It is not a case of food-borne illness,[4] however, since there is no evidence that it is transmitted through food (other than eating the alleged wild animals). It is also possible that it was not food-based, through eating the animal. The proximity of humans and animals, including wild animals, in the Huanan Market allows for a zoonotic spill over but the exact route has not been ascertained. Transmission can be through various routes. Eating following cooking may not be dangerous if the cooking has killed off the virus. Avian influenza, another novel coronavirus, was transmitted from birds to humans by touch, rather than by eating.[5] There was little evidence of any onwards transmission from human to human (World Health Organization, 2012[10]). Other diseases, such as campylobacter, spread from animals by faecal matter contaminating raw food, including animal carcasses at slaughter (World Health Organization, 2020[11]).

There is no evidence of COVID-19 being food-borne and the WHO asserts that "There is currently no evidence that people can catch COVID-19 from food or food packaging" (World Health Organization, 2020[12]). In September 2020, the International Commission on Microbiological Specifications for Foods' (ICMSF) delivered its final opinion on SARS-CoV-2 and its relationship to food safety, affirming the unlikeliness of the virus being a food risk (Box 1.1), in the sense of being transmitted through food. The assertion regarding food packaging may be more contentious since there have been claims that the virus has spread on frozen packaging of seafood, but these claims have been challenged (Heidt A, 2020[13]). Even if these claims were true, the transmission is still not through the food itself but through frozen packaging that happens to contain food. Eating the frozen food that had been shipped would not have transmitted the virus. The food industry played an additional role in the spread of the disease insofar as many slaughterhouses became "superspreader" sources of transmission. This was attributed to the cold and damp conditions inside, the prevalence of metal surfaces, the proximity of workers and the tendency for them to shout above the normal noise (therefore increasing aerosol transmission). Some also had cramped living conditions for the workforce (Middleton, Reintjes and Lopes, 2020[14]). But there was no evidence that this led to any transmission through the food coming from these establishments, although that has also been caught up in the debate on whether the virus can be transmitted in frozen food or food packaging (Fisher D et al., 2020[15]).

Box 1.1. The ICMSF's opinion on SARS-CoV-2 and its relationship to food safety

The International Commission on Microbiological Specifications for Foods' (ICMSF) opinion on SARS-CoV-2 and its relationship to food safety of 3 September 2020 states as follows:

"SARS-CoV-2 should not be considered a food safety hazard since a true food safety hazard enters the human body with food via the gastro-intestinal (GI) tract, where it can infect organs/tissues elsewhere in the human body." "Despite the many billions of meals consumed and food packages handled since the beginning of the COVID-19 pandemic, to date there has not been any evidence that food, food packaging or food handling is a source or important transmission route for SARS-CoV-2 resulting in COVID-19".

Source: (ICMSF, 2020[16]).

The novel virus mutated in the genetic animal reservoir and jumped the species to humans. Later into the pandemic, a mutated form of the virus was found in farmed minks but It has not been found in farmed animals, so it is more likely to have come from wild animals, whether alive or dead, such as bats. It could have transferred to an intermediary animal before transferring to humans. It is very well adapted to human cell receptors, which enables it to invade human cells and easily infect people. Other viruses, including others in the coronavirus family, may also have transmitted from animals to humans but had no appreciable effect so the simple fact of a zoonotic jump is not necessarily unusual. It is also a virus that transmits easily from human to human, unlike the Avian Influenza coronavirus, and allows for asymptomatic transmission, which is a very potent combination. The final factor which made this zoonotic transmission into a pandemic was the proximity to an international airport serving a large population. The Ebola virus did not convert to a pandemic on the scale of COVID-19 partly through its outbreaks occurring in less populated areas, away from highly-trafficked air routes, although it was highly infectious human-to-human. It was very lethal and very visible, unlike COVID-19 which has a high proportion of asymptomatic and "light" cases. Of all the factors in the causal chain between mutation of the virus within its genetic reservoir and the deaths of over a million people globally, the standard of food hygiene in the Huanan Market was at most a necessary but insufficient condition. That is, a higher standard of food hygiene might have prevented that causal chain, but a pandemic might still have resulted from another route. Not only is there still uncertainty about the exact sequence of events, but there is also uncertainty about whether it could have happened again elsewhere. Given the spread of the disease, further direct transmission to humans may make little difference but the Huanan Market was closed down and disinfected on 1 January 2020 and remains closed. Other markets in Wuhan reopened after the city's shutdown ended (ABC News, 2020[17]). There still remains an argument for improving food hygiene levels in wet markets but that argument has been widely accepted for some time. The difficulty is how to achieve it.

It should also be mentioned that the whole narrative of the Huanan Market being the source of the virus transmission has also been challenged (Cohen J, 2020[18]) (Letzter R and Writer S, 2020[19]). It took 15 years to settle the question of the source of the SARS virus, so it may take many years yet to determine the exact source of COVID-19. What does seem to be reliable is the role of that market in amplifying the spread of the virus. Whether or not the virus was transmitted for the first time from animals to humans through some action in that market, there is much clearer evidence that many of the early human cases of infection were traced back to the market (Mizumoto, Kagaya and Chowell, 2020[20]). That could be a feature of a crowded enclosed space which happened to be a food market or there may be other factors related to being a food market that made it more likely that it would amplify transmission, such as diversity of attendees from different districts or a greater than usual number of common touch surfaces.

The "One Health" approach and its potential to contribute to pandemic risk management

The previous chapter showed that COVID-19 was not necessarily a food safety issue and had greater trappings of a public health issue – but that does not mean that the food chain has no relevance in pandemic risk mitigation. Although it may have its origin in the food chain, insofar as it may have been transmitted to humans through eating or the proximity with animals gathered for food sale, the transmission may have had nothing to do with eating. There is no evidence that it is a food-borne illness, i.e. that it is capable of being transmitted through food, and such evidence is bound to have emerged within a short time after the outbreak. The connection with food-borne illness is only though packaging of food and whether even that is the case is contentious. The source of the pandemic may have been a food market but even that has not been fully ascertained and may not be for years. What can be asserted is that a particular food market had an amplifying effect on transmission but that may not have been because of anything exceptional about it being a food market. So, it is far from clear from this case that preventing the next pandemic is a food safety risk that should be managed as such.

This, however, may be taking too narrow a view of the scope of the topic of food safety, and not considering the broader impact of the food chain on health, and specifically on zoonotic and pandemic risks. At its narrowest, food safety is about protecting consumers against unsafe food. However, to achieve that can stretch the scope to an unmanageable extent. Eating is fundamental to human life and therefore a large part of human resource and activity goes into providing food. It is one of the most basic markets in any society, as well as one of the most global in the present day. Food that kills or sickens defeats the purpose so ensuring the safety of food is a basic task within the activity of providing food, whether production, distribution, catering or even domestic cooking. Preventing zoonotic transfer of a dangerous virus into the food supply can clearly come within that scope but the issue is less one of the outcome to be achieved as the scope of what needs to be done to achieve it. The fact that some change may result in safer food is not sufficient to bring that activity within the regulatory remit of food safety. Reducing world poverty will result in safer food but that does not extend the food safety regulatory remit to include anything that reduces world poverty.

To illustrate the relevance of food safety to many human activities, Box 1.2 lists the number of the Sustainable Development Goals (SDGs) that are affected by actions taken as part of ensuring food safety. That does not mean that all these SDGs are part of the regulatory remit of food safety but, just as efforts to deliver food safety will assist in achieving these SDGs, working towards these SDGs may also improve food safety through improving the context in which the food safety factors operate. It is this bi-directional interaction that is the complicating factor (including the complication that the interaction may sometimes be negative). Scientific research into pathogens or bacteria that may reside in food are direct applications of food safety for its own sake whereas changing how humans relate to animals in order to reduce zoonotic disease go much wider than efforts to improve food safety, even though food safety may improve as a result.

Box 1.2. Food safety and the Sustainable Development Goals

Food safety will be vital for achieving many of the Sustainable Development Goals (SDGs), and particularly the following:

- SDG 1: End poverty. Foodborne disease (FBD) is a major cause of ill-health among the poor and is associated with a range of costs affecting them, including lost workdays, out-of-pocket expenses, and reduced value of livestock and other assets.

- SDG 2: End hunger. FBD has multiple complex interactions with nutrition. For example, toxins may directly lead to malnutrition, some of the most nutritious foods are the most implicated in FBD, and concerns over food safety may lead consumers to shift consumption away from nutritious foods.

- SDG 3: Good health and well-being. The health burden of FBD is comparable to that of malaria, HIV/AIDS, and tuberculosis, and the people most vulnerable to FBD are infants, pregnant women, the elderly, and those with compromised immunity.

- SDG 5: Gender equality. Women are the gatekeepers of household food safety, play important roles in traditional food chains, and often derive their livelihood in agri-food value chains.

- SDG 6: Clean water and sanitation. Lack of clean water increases the risk of food being unsafe, injudicious use of chemicals in food production can pollute water sources, and infectious FBDs can be transmitted via water.

- SDG 8: Decent work and economic growth. Inclusive food markets provide livelihoods and are a way out of poverty for many poor people.

- SDG 11: Sustainable cities and communities. Hundreds of millions of poor people work in urban agriculture and food-related services, and vibrant traditional food markets and street food make important contributions to culture, tourism, and liveable cities.

Source: (Jaffee et al., 2019[21]).

At its narrowest, food safety is about preventing "unsafe" food, whether from contamination or even deliberate adulteration by businesses. But even with that narrow definition, a wild animal carrying a dangerous virus would constitute "unsafe" food, if it could be regarded as "food". Following the concern over the Wuhan outbreak, various Chinese cities banned the eating and, in some cases, farming of wild animals. Yet, in many countries owing to socio-economic factors, food safety standards do not apply to wild meat. Food safety regulations become more complicated as details are considered. The concept of "wild" animal is too vague. Some religions ban the eating of specific animals, with pig meat being the most familiar, and there are also strong cultural pressures about eating particular species which are happily eaten in other cultures, such as horses or dogs. But contrasting "wild" with "farmed" is more helpful and gives a more rational foundation on the basis that farming allows some level of control of the animal as a source of food. Banning the farming of wild animals therefore seems a contradiction, although it may be a matter of degree.[6] Banning the trade in "wild" animals avoids the complication of farming but there are already bans on the trade in many animals, primarily under the CITES Convention, although that focuses on endangered species (CITES, 2021[22]). Following the COVID-19 outbreak, some Chinese authorities are now trying more targeted regulation based on species or a range of activities related to interaction with various species.[7]

An alternative approach to the problem of zoonotic disease has operated since the early 2000's, under the heading of "One Health". It is a collaborative, multisectoral, and transdisciplinary approach – working at the local, regional, national, and global levels – with the goal of achieving optimal health outcomes recognizing the interconnection between people, animals, plants, and their shared environment (Centers for Disease Control and Prevention, 2018[23]). The animal element is actively supported by the World Organisation for Animal Health (OIE), and their website gives a clear argument based on zoonotic disease, summarised in Figure 1.1 below (World Organisation for Animal Health, 2021[24]).

Figure 1.1. Impact of zoonoses

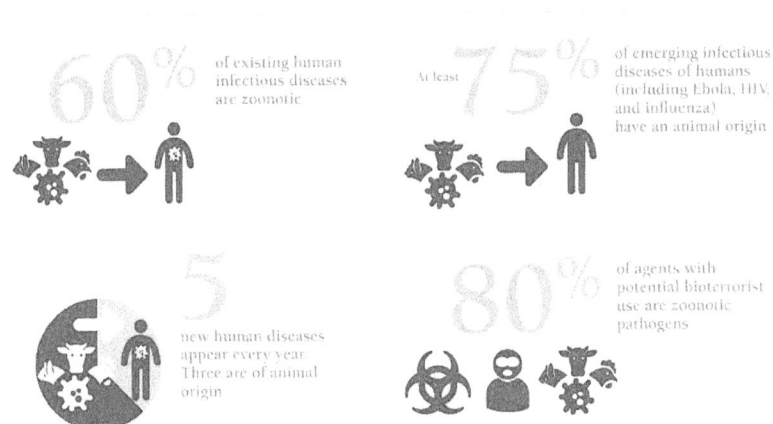

Source: (World Organisation for Animal Health, 2021[24]).

The WHO also supports the One Health approach, partly for zoonotic disease but additionally for the new danger of antimicrobial resistance,[8] which it sees as another interface between human and animal worlds (World Health Organization, 2017[25]). The US Centers for Disease Control and Prevention (CDC) take it even further in looking at environmental changes and demographic shifts which alter the physical interface between humans and animals as habitats change and more humans have greater exposure to wild animals. This is a logical extension, but it illustrates how easily one subject can expand. One Health is not a specific discipline, like food safety, but rather an approach to take in a collection of disciplines, to be aware of each discipline's interaction and effects on others. As regards food safety, the OIE diagram at Figure 1.1 refers to diseases, rather than food-borne illness or food-related illness, therefore goes much wider. But the diseases are results of pathogens which may also affect food. For food safety, the issue is the generating of these pathogens in animal gene pools and this has been fundamental to the study and practice of food safety. Foods of animal origin always carry a higher risk than all other sources of food.

The pandemic has focused concern yet again towards the practices that involve interaction between humans and both farmed and wild animals, dead and alive, which are considered to have been part of the origin of earlier virus outbreaks, including Ebola, SARS, MERS and Avian Influenza. That can include methods of animal husbandry and also the operation of wet markets, small and large. Disruptions in the food chain, such as the extensive damage to the pork value chain due to an unprecedented spread of African Swine Fever in China (Patton D, 2021[26]), may also contribute to the complexity of the issue insofar as alternative sources of protein may be sought. It could be argued that this is just in the nature of zoonotic disease, i.e. spillover of a virus across species, and that it is other factors such as concentration of population and ease of global transport that has transformed traditional practices into deadly pandemics. But these practices remain part of the issue and being traditional is not a compelling reason to let them continue. This is what the multidisciplinary approach of One Health is designed to deal with. These issues are connected with the food production system even if they are not food-borne illnesses and therefore those involved in food safety, including the food industry, need to be active members of One Health and need to take responsibility for whatever their contribution to the interlocking issues may be. Zoonotic disease control is a genuinely complex issue and merely being a contributor to that complexity matters.

While some regulations including bio-security measures already exist for farmed animals, the issue of zoonotic disease and food safety can be presented as whether food safety regulation should try to regulate to prevent zoonotic diseases. Some of the complexities of this were covered when considering the various regulatory responses in Chinese cities to the outbreak attributed to the Wuhan market. But they are not an integral part of the Chinese food safety regulatory regime but rather the regulation of the trade in animals and so implementation is unlikely to be through the local government system for food safety (Xie E, 2020[27]). Regulation by the state is certainly an element in action to tackle the sort of issues covered by One Health but that needs to be divided across specific regulatory regimes. That will also bring with it considerations of the structure of governments since regulation will be the responsibility of particular ministries and this will vary from one country to another. A complex issue like managing the risk of further pandemics will have a regulatory element but it will involve a mixture of ministries and regulatory regimes, just as One Health is a multidisciplinary movement.

Managing the problem of wet markets

Wet markets are another key element in the analysis of how the constant threat of zoonotic disease may result in pandemics. But these are themselves complex socio-economic problems and go beyond simple labelling as food safety. Although the centre of food safety is the issue of pathogens in otherwise safe food, much of the practice relates to human behaviours in managing the risks presented by these pathogens. The substance carrying the pathogen only becomes food when it enters the food chain. Food chains can become extremely complicated and lengthy, and they will often have bottlenecks, where the risk of mismanagement of the food increases. One of the main bottlenecks is markets. They are a pinch point where a host of risks intermix, from the number of people, business interests, animals, food products, pests and pathogens that interact in a confined space, with many hygiene challenges. For animals to enter the food chain, they have to be slaughtered and that can create another bottleneck. But some markets are also where animals are killed on the spot or slaughtered locally. "Wet" markets are where fresh meat, fish, living animals or other perishable goods are sold, in contrast to "dry" markets which sell textiles and hardware. Wet markets therefore present a major challenge to food safety.

However, there is no easy answer to the problem of wet markets. They can be regulated but that in itself does not solve the problem since implementation may be ineffective and wider socio-economic factors may further complicate the issues. In developing economies, many wet markets will also be in the informal sector, i.e. outside the operation of the regulatory system, but the weaknesses in regulatory implementation within the formal sector may make the difference between formal and informal less of an issue. What is more important than the regulatory issues are the socio-economic drivers. Wet markets are fundamental to the local economies and also to the social functioning of the local communities. The customers are often in the poorer segment of the market and that gives them specific demands, such as small portions of fresh food for immediate consumption. A study in Southern Africa showed that informal traders could buy food products from supermarkets and make a profit selling them in local markets because they had divided them into smaller portions which were more affordable, although more expensive by weight (Crush J and Frayne B, 2018[28]). This is one illustration of the debate over the last ten years in development economics (and food safety) around "supermarketisation".

In summary, "supermarketisation" is a debate between the roles of government and of the private sector in driving development of food value chains, including production and also retail. The term comes from the advent of supermarkets as part of urbanisation, belonging to national or even global chains. These could be relatively small "convenience stores" or hypermarkets but would be driven by private sector. Some companies would also create or support much of the value chain from initial production through to distribution and retail by various methods, including creating their own farms. This activity by the private sector can significantly affect the agri-business sector and the distribution and retail sectors. The debate arises where that development replaces or delays deeper and wider development of these sectors by government action.

There are strong advocates of supermarkets driving development in Lower- and Middle-Income Countries (LMICs). Indeed, it remains a policy of the Vietnamese government in trying to modernise its agricultural, rural and urban economies (Wertheim-Heck, Vellema and Spaargaren, 2015[29]). Advocates argue that it leads to higher standards of food safety because it is backed by the global food companies rather than national governments. These companies can create their own supply chains in the country by working as vertical integrators, providing a reliable market for small producers which allows them to build stronger businesses. There is a wide variety of methods through which global food companies can create what can amount to a micro economy or even be regarded as a micro regulatory environment.[9] Some variations are:

- The company supplies piglets to householders, plus feed, and guarantees a purchase price on maturity: this can strengthen livestock breeding in terms of genetic stock and initial health of the piglet.
- The company provides smallholdings, including accommodation, plus all agricultural inputs and training, and markets the resulting products.
- The company enters into a contract with collectives which gives it shared rights to land for farming and it manages diversification across the collective, in addition to providing a guaranteed market for the produce.

One illustration of this comes from the Metro supermarket chain, operating through "Star Farm" networks in both China and Pakistan. A key element is providing a traceability system for the food products in its supermarkets.[10] Customers can use a smartphone app to scan a barcode to display the entire history of the product, which gives them some reassurance about the safety of the food.

Opponents argue that this leads to capture of the rural economy by the elite, while covering a small percentage of the overall rural economy and giving the government cover to delay or avoid wider development strategies (Jaffee et al., 2019[30]). Food safety standards will generally be higher in supermarkets than in informal markets but that is not always guaranteed, and even informal markets need to ensure a certain level of food safety – they cannot survive if they seriously poison their customers through poor food hygiene. It is also argued that the food is often more nutritious in informal markets than the ultra-processed food that is the staple in many supermarkets. But the arguments in favour of traditional wet markets are primarily socio-economic. One study of a wet market and slaughterhouse in Ibadan, Nigeria, noted that a modern market and slaughterhouse built to replace the unhygienic old market was rejected by traders, with some even being killed in rioting to stay where they were (Grace, Dipeolu and Alonso, 2019[31]). As explained in that article:

> "The case study also provides an example of how attempts to upgrade value chains can be problematic if they do not take into account the context and the complexities of governance. The modern abattoir had objectively better facilities, but the location was less convenient and the costs for the butchers higher. It was not apparent that any market survey had been carried out to establish demand. There are many other examples in developing countries where modernisation of infrastructure resulted in facilities that were less acceptable to traders and customers. For example, a well-documented case from Lusaka examined how street vendors were moved into new and hygienic premises. However, most returned to their former positions as the improved market was less accessible to customers and entailed more transaction costs for traders, even though the environmental conditions were better (Ndhlovu PK, 2011[32])".

The International Livestock Research Institute (ILRI) is one of the leading supporters of informal wet markets, including the role of these markets in the recent growth of livestock as a food in LMICs.[11] The related International Food Policy Research Institute (IFPRI) has also published a defence of wet markets specifically in China (Alita L, 2020[33]) but it has also published a blog arguing against the "either/or" mindset in the supermarketisation debate (Ranieri J and Wertheim-Heck S, 2020[34]), with various suggestions for more varied policies to support the smaller businesses:

- Identifying innovative food safety policies and interventions, such as participatory guarantee systems, to improve wet market vendor hygiene and food handling practices.
- Implementing low-cost safety control mechanisms and policies to renovate and upgrade existing traditional fresh food outlets, improve business standards, and offer an alternative to closure.
- Removing supermarket access barriers for poorer consumers: in time, convenience, cost, and perceptions, without jeopardizing diet quality.
- Creating more effective in-store food quality control and consumer awareness campaigns to improve trust in food safety guarantees and education about diet and health risks associated with ultra-processed foods, and the importance of continuing to eat fresh produce.

What this illustrates is the importance of placing food safety regulation within a wider policy context than simply food hygiene. The supermarketisation debate is primarily related to LMICs whereas the role of supermarkets and indeed the global food companies is very different in developed economies (Shucksmith and Brown, 2016[35]). The approach to food safety in LMICs within a wider economic perspective has recently been usefully summarised in an article in World Development journal (Hoffmann, Moser and Saak, 2019[36]). The World Bank Group (Jaffee et al., 2019[30]) takes this further in positing a "food safety lifecycle" that developing economies go through as they transit from Low Income to Upper Middle Income. Designing food safety regulatory frameworks needs to be tailored to these contexts, rather than following one regulatory model or one economic model.

Delivering food safety

Food safety is delivered by what people who manage food do with it. The levers that determine what they do with it include market forces, regulations, habits, information and awareness, social pressures and even just how they feel at the time. Regulations are only one of these levers, and not necessarily a strong one. The regulations may be specifically part of a food safety regulatory regime, but they may also act on other levers, especially on market forces, as seen in the previous discussion about wet markets, or on animal husbandry, as seen in the One Health discussion. But in each case the regulations are only a tool for government to try to influence behaviour, as opposed to a lever that has a direct (and intended) consequence. Good regulatory design matters but what matters equally is how regulations are implemented and scoring the right balance between market forces and regulations.

An important factor in food safety regulation is that it is usually accompanied by an implementation infrastructure, rather than just left on the books in the legislature or jumbled with numerous other regimes and given to local governments to manage somehow. There will often be an institution dedicated to delivering the regulatory outcomes, with research and analysis capabilities and direct or indirect access to a network of people on the ground to take action. Where such an institution is focused on the regulatory outcomes, rather than just the technicalities of specific regulations, it becomes an important actor in the complexity of delivering food safety. It is that actor who can connect through One Health with other agencies within the multidisciplinary approach to find collaborative solutions for its part in the complex challenge of managing zoonotic disease.

In general, the traditional approach to regulation was "command and control" or "enact and enforce". It assumed that delivery of the regulatory outcomes depended on stopping bad things happening. Especially with food safety but also with other safety-based regulatory regimes, there is now a recognition that the regulatory outcomes require proactive, preventive, positive action. This is a new paradigm but now well established across developed economies. The force of law may be needed to make people do things against their will but delivering better ways of working and higher practice standards does not require enforcement. It requires new techniques from regulatory agencies, primarily in engaging with the regulated subjects and enabling them to improve. The paradigm shift is from "enact and enforce" to "engage and enable".

One food safety regulator serves as a relevant example about how delivery of regulatory objectives has transformed to become more engaging and enabling. The Food Safety and Standards Authority for India (FSSAI) has taken a radical approach to its delivery role. Its remit under the Food Standards and Safety Act 2006, Section 18, is – *"to ensure safe and wholesome food for human consumption"*. The addition of "wholesome" allowed it to cover both safety and nutrition but it then went a stage further and built on its powers in relation to packaging to claim authority to implement some environmental element of a food systems approach. The result is a programme called "Eat Right India" which promotes the following philosophy (FSSAI, 2021[37]): "If it's not safe, it's not food; If it's not healthy, it's not food; and if it's not good for the planet, it's not food."

It sees its remit as widely as that, but its exercise of that remit is also radical. It has a graded approach to delivering its objectives.

- **For large food businesses**: FSSAI uses traditional regulatory instruments and tools with a focus on schemes of testing and inspections. Its banning of Maggi Noodles in 2015 demonstrated that it was also a robust enforcer where appropriate (Chhabara Rajesh, 2015[38]).
- **For small and medium food businesses**: The focus is largely on capacity building and hygiene ratings to promote self-compliance. The purpose is to improve hygiene conditions especially at the manufacturer's level. Under the authority's hygiene rating initiative, food businesses are given a rating (five to one) to reflect the level of hygiene and food safety compliance, based on the English Food Hygiene Rating System.
- **For micro and informal businesses**: it applies a "cluster approach" – a systematic process of gap analysis, filling infrastructure gaps, training and capacity building and certification. It works with clusters of hawkers, up to one street at a time, rather than individuals. They have to register with FSSAI but this does not in itself take them into the formal sector as individual businesses.

There is much to learn from the FSSAI approach to delivering food safety. A conventional inspection-based approach would not go far in tackling the scale of the problem. It is trying to ensure the supply of safe and wholesome food to 1.3 billion people daily, equivalent to 16.66% of the entire world population. It sees its main task as raising practice standards rather than sanctioning non-compliance. Closing down a bad business does not create a good business and India needs millions of good businesses in order to meet the objective. Urbanisation is putting even more stress on street food and it is resisting supermarketisation, which means that FSSAI has to deal with millions of micro businesses in the informal sector. Rather than "enforce", its mantra is "engage, excite, enable". So, it promotes campaigns that provide information and motivation to improve food hygiene. It partners with many other bodies and builds new networks in order to expand its reach. It has produced more YouTube videos than secondary regulations.

This approach not only fits with the expansive view of food safety as a topic but also aligns with thinking on how to deliver real progress in managing food. There is a common saying in the development profession that "you can't regulate your way to food safety", mainly because the majority of transactions in LMICs are in the informal sector which is, by definition, unregulated. Instead, the focus is on an approach referred to as TCM – Training, Certification and Marketing (Grace, Dipeolu and Alonso, 2019[31]). Training is essential if standards are to rise but it is insufficient. Something more has to result from it, which is often some form of certification or recognition that the business has improved (which can fit with the hygiene rating schemes). It also needs motivation because commercial pressures will not always align with improved standards.

One useful example from the Eat Right India range of interventions and campaigns is Clean Street Food Hub. This aims to improve standards in street food and applies the FSSAI approach. Groups of traders for an area are registered and trained but they are also enabled. The local government provides clean water, electricity and waste disposal services for the traders, without which the training would achieve little. FSSAI is now working with local governments across India to provide that infrastructure as part of urban planning, rather than as a one-off. It is funding 500 Clean Street Food Hubs but challenging the State authorities to expand that to 15 000 (FSSAI, 2021[39]). The approach also meets the TCM model.

It was not the first to recognise that engagement with the regulated businesses had to be effective and not formal. In 2007, the United Kingdom Food Standards Agency (FSA) faced the challenge of new EU regulations that required even small food businesses to apply Hazard Analysis Critical Control Point (HACCP) principles, a method of managing food safety (and other) risks but it can be very challenging and expensive to implement in full. It produced a pack for small family-run restaurants called "Safe Food, Better Business" which consisted of a lever-arch folder with wipe-clean sheets of instructions on basic operations in a small restaurant, using pictures and diagrams with minimal text. It also had a DVD in 14 languages, to cover some of London's diversity of national cuisines. It was a seminal approach that has been followed or imitated in many countries, with even a Mongolian edition released in 2016. A study by the Chartered Institute of Environmental Health[12] also showed that 68% of the businesses who used it also improved their commercial business by applying a more systematic approach to management of food products, and thus reducing waste etc.

This latter point is also an important insight to the new approach to regulatory delivery – not that regulatory intervention should always improve business but that improving business was a relevant consideration. A very common approach to implementing food safety regulation has been the use of an official inspection result as a marketing point to attract consumers. It started first in San Francisco late last century but can be found in many countries across the globe, including Nepal. It is interesting that this technique is almost exclusive to food safety regulation and has not been successfully applied in other regimes.

Conclusions

Delivering the objectives of a regulatory regime for food safety does not depend on a relationship with business based on sanctions. Regulation may be needed to empower or enable, and it is also important in shaping what good looks like, through reference to standards. But to be effective it needs a delivery agency to provide organisation, focus and also a contact point for other bodies with related objectives. A modern regulatory delivery agency has to work beyond just assuring compliance with technical requirements. It needs to understand the sector it is regulating and be aware of what changes the sector may need that come from other parts of government[13] as well as modifications needed in its own regime. The relationship with the businesses in the sector has to be one of engagement rather than top-down supervision (while avoiding "regulatory capture"). This can be seen in the way that guidance in relation to new regulations has moved from lengthy texts that largely repeat the regulations to infographics, videos and social media. Most modern regulatory delivery agencies will have a well maintained Facebook page. The application of the new discipline of behavioural insights also illustrates how far "enforce" has become "engage".

That sort of delivery agency is a way for government to overcome the constraints of organisational silos, by having teams that can collaborate with other parts of government and external partners through the various multidisciplinary networks that have been referred to earlier. Food safety regulation and food safety agencies do not have to regulate trade in animals in order to be an active part of the One Health network along with the specialists who are better placed to guard against further pandemics arising from another zoonotic spillover. The importance of policy coherence across related areas of regulation has been stressed in the OECD Recommendation on Policy Coherence for Sustainable Development (PCSD) (OECD, 2019[40]) and is actively pursued by OECD.

But perhaps the biggest change in the delivery of safer food is the recognition that techniques for engagement and enabling can transcend the regulatory regime. They can be applied in informal markets and not just to registered businesses. FSSAI cannot make much headway in the scale of its challenge without finding a way to improve practice levels in the informal economy. When looking at ways of preventing future pandemics by changing behaviours, that will have to include changing behaviours in the informal economy as well.

Notes

[1] SDG 7 deals with affordable and clean energy. SDG 13 and 14 deal with climate action and life below water respectively.

[2] For a short summary of the "supermarketisation" debate, see https://www.ifpri.org/blog/supermarketization-food-environments-and-urban-poor.

[3] Overview of most widespread zoonoses, https://absa.org/wp-content/uploads/2017/01/zoonoticfactsheet.pdf.

[4] For an explanation of what constitutes food-borne illness, see http://www.fao.org/fao-who-codexalimentarius/sh-proxy/en/?lnk=1&url=https%253a%252f%252fworkspace.fao.org%252fsites%252fcodex%252fmeetings%252fcx-712-51%252fcrd%252ffh51_crd06x.pdf.

[5] See a CDC infographic on transmission of avian flu, https://www.cdc.gov/flu/pdf/avianflu/avian-flu-transmission.pdf.

[6] This New York Times article illustrates this quandary since the subject has been farming bamboo rats for five years so may have progressed from simply containment and breeding to a form of quality control if it is a sustainable food business, https://www.nytimes.com/2020/06/07/world/asia/china-coronavirus-wildlife-ban.html.

[7] This South China Morning Post article summarises a range of regulatory responses, https://www.scmp.com/news/china/society/article/3085467/coronavirus-wuhan-confirms-chinas-ban-trade-eating-wild-animals.

[8] This is also the case for bacterial infections. Most food-borne pathogens are bacterial in origin. Antibiotic resistance is increasingly becoming a cause of international concern. Some countries such as

Sweden in their strategy to combat antibiotic resistance are setting an example for policy coherence. https://www.government.se/articles/2020/04/updated-swedish-strategy-to-combat-antibiotic-resistance/.

9 For an insight into the world of "contract farming", see the FAO / Unidroit / IFAD "legal Guide to Contract Farming" Rome 2015, https://www.unidroit.org/studies/contract-farming.

10 https://www.metro.cn/en/metro-food-safety/traceability.

11 See this blog: https://news.ilri.org/2015/01/27/despite-contamination-concerns-africa-must-embrace-wet-markets-as-key-to-food-security/ from 2015, which ILRI has put back on its landing page in 2020, and the book that it is based on: https://cgspace.cgiar.org/bitstream/handle/10568/42438/Food%20Safety%20and%20Informal%20Markets.pdf?sequence=4&isAllowed=y.

12 Available at : https://web.archive.org/web/20071020072540/http://www.cieh.org/library/Knowledge/Food_safety_and_hygiene/Case_studies/Westminster%20CHIP.pdf.

13 This is similar to the "Regulatory Stewardship" model that has applied in New Zealand since 2013, https://www.treasury.govt.nz/information-and-services/regulation/regulatory-stewardship.

References

ABC News (2020), "Wuhan's 'Wet Markets' Are Back in Business", *ABC News.* [17]

Alita L (2020), "Supermarketization is not the Only Strategy to Enhance Food Safety in China", *Agriculture for Nutrition and Health.* [33]

BBC News (2020), "Coronavirus Was Already in Italy by December, Waste Water Study Finds", *BBC News.* [9]

Blanc, F. and D. Macrae (2021), "Food Safety Compliance", in *The Cambridge Handbook of Compliance.* [4]

Centers for Disease Control and Prevention (2018), *One Health Basics.* [23]

Chhabara Rajesh (2015), *Asia column: Lessons from Nestlé's crisis in India*, Reuters, https://www.reutersevents.com/sustainability/stakeholder-engagement/asia-column-lessons-nestles-crisis-india (accessed on 1 March 2021). [38]

CITES (2021), *What is CITES?.* [22]

Cohen J (2020), *Wuhan Seafood Market May Not be Source of Novel Virus Spreading Globally.* [18]

Fisher D et al. (2020), *Seeding of outbreaks of COVID-19 by contaminated fresh and frozen food.* [15]

FSSAI (2021), *Clean Street Food Hub*, https://fssai.gov.in/CleanStreetFood/indexhome (accessed on 1 March 2021). [39]

FSSAI (2021), *Eat Right India*, FSSA, https://eatrightindia.gov.in/EatRightIndia/index.jsp (accessed on 1 March 2021). [37]

Grace, D., M. Dipeolu and S. Alonso (2019), "Improving Food Safety in the Informal Sector: Nine Years Later", *Infection Ecology & Epidemiology*, Vol. 9/1, http://dx.doi.org/10.1080/20008686.2019.1579613. [31]

Heidt A (2020), "Coronavirus Found on Food Packaging, but Likely of Little Concern", *The Scientist*. [13]

Hoffmann, V., C. Moser and A. Saak (2019), "Food Safety in Low and Middle-Income Countries: The Evidence Through an Economic Lens", *World Development*, Vol. 123, http://dx.doi.org/10.1016/j.worlddev.2019.104611. [36]

ICMSF (2020), *ICMSF opinion on SARS-CoV-2 and its relationship to food safety*, https://www.icmsf.org/wp-content/uploads/2020/09/ICMSF2020-Letterhead-COVID-19-opinion-final-03-Sept-2020.BF_.pdf (accessed on 2 March 2021). [16]

Jaffee, S. et al. (2019), *The Safe Food Imperative: Accelerating Progress in Low- and Middle-Income Countries*, Washington, DC: World Bank, http://dx.doi.org/10.1596/978-1-4648-1345-0. [21]

Jaffee, S. et al. (2019), *The Safe Food Imperative: Accelerating Progress in Low- and Middle-Income Countries*, Washington, DC: World Bank, http://dx.doi.org/10.1596/978-1-4648-1345-0. [30]

Letzter R and Writer S (2020), "The Coronavirus Didn't Really Start at that Wuhan 'wet market'", *Live Science*. [19]

Middleton, J., R. Reintjes and H. Lopes (2020), "Meat plants—a new front line in the covid-19 pandemic", *BMJ*, p. m2716, http://dx.doi.org/10.1136/bmj.m2716. [14]

Mizumoto, K., K. Kagaya and G. Chowell (2020), "Effect of a wet market on coronavirus disease (COVID-19) transmission dynamics in China, 2019–2020", *International Journal of Infectious Diseases*, Vol. 97, http://dx.doi.org/10.1016/j.ijid.2020.05.091. [20]

Ndhlovu PK (2011), *Street Vending in Zambia: A case of Lusaka District*. [32]

OECD (2021), *Making Better Policies for Food Systems*, OECD Publishing, Paris, https://dx.doi.org/10.1787/ddfba4de-en. [3]

OECD (2019), *Recommendation of the Council on Policy Coherence for Sustainable Development*, https://www.oecd.org/gov/pcsd/oecd-recommendation-on-policy-coherence-for-sustainable-development.htm. [40]

Parodi E (2020), "Researchers say study on COVID-19 in Italy doesn't dispute virus origins". [8]

Patton D (2021), "New China Swine Fever Strains Point to Unlicensed Vaccines", *Reuters*. [26]

Ranieri J and Wertheim-Heck S (2020), "'Supermarketization,' food environments, and the urban poor", *International Food Policy Research Institute*. [34]

Routledge (ed.) (2018), *The 'Supermarketization' of Food Supply and Retail*. [28]

Shucksmith, M. and D. Brown (eds.) (2016), *Routledge International Handbook of Rural Studies*, Routledge, Abingdon, Oxon ; New York, NY : Routledge, [2016], http://dx.doi.org/10.4324/9781315753041. [35]

Wacharapluesadee, S. et al. (2021), "Evidence for SARS-CoV-2 related coronaviruses circulating in bats and pangolins in Southeast Asia", *Nature Communications*, Vol. 12/1, http://dx.doi.org/10.1038/s41467-021-21240-1. [2]

Wertheim-Heck, S., S. Vellema and G. Spaargaren (2015), "Food safety and urban food markets in Vietnam: The need for flexible and customized retail modernization policies", *Food Policy*, Vol. 54, http://dx.doi.org/10.1016/j.foodpol.2015.05.002. [29]

Wertheim, J. et al. (2013), "A Case for the Ancient Origin of Coronaviruses", *Journal of Virology*, Vol. 87/12, http://dx.doi.org/10.1128/JVI.03273-12. [5]

World Health Organization (2020), *Campylobacter*. [11]

World Health Organization (2020), *Coronavirus disease (COVID-19): Food Safety for Consumers*. [12]

World Health Organization (2020), "How WHO is working to track down the animal reservoir of the SARS-CoV-2 virus". [7]

World Health Organization (2020), *Origin of SARS-CoV-2*, WHO. [6]

World Health Organization (2017), *One Health*. [25]

World Health Organization (2012), "Influenza: H5N1". [10]

World Organisation for Animal Health (2021), *One Health "at a Glance"*. [24]

Xie E (2020), "China Bans Trade, Eating of Wild Animals in Battle Against Coronavirus", *South China Morning Post*. [27]

Zarocostas, J. (2021), "WHO team begins COVID-19 origin investigation", *The Lancet*, Vol. 397/10273, p. 459, http://dx.doi.org/10.1016/s0140-6736(21)00295-6. [1]

2 The "culture of food safety" as a model to make behaviour safer

This chapter considers the challenges posed by widespread transformation of behaviour to improve safety (particularly in terms of health and epidemic control), seeking to build on the experience of the "culture of food safety" and how it has profoundly transformed practices over the past couple of decades. Difficulties in achieving compliance with COVID-19 safety measures has shown the urgency of going beyond mere rule-setting and formal enforcement, and designing strategies and programmes to promote and achieve safer conduct at work and in social life. Understanding how culture change and safety culture have been systematically built up in the food sector through a combination of methods, systems, management and regulatory attention, can provide an important contribution.

Introduction

As noted regulation scholar Robert Baldwin once remarked, "rules don't work" (Baldwin, 1990[1]). This need not mean that rules are not useful or important, but that they do not "work" (produce results) in and of themselves. The importance of transforming rules into practice has been emphasised repeatedly, leading to the emergence of the term "regulatory delivery" (Russell, 2019[2]). Increasing attention has been given to how behaviour (and, specifically, behaviour of managers and employees in business organisations) is driven largely by "culture" – and how important it is for regulatory systems to try and understand, harness, and influence corporate culture (Hodges, 2015[3]) (Hodges, 2018[4]). The OECD has looked into the impact of behavioural approaches on safety (OECD, 2020[5]), which is an important part of such work. Here, we look specifically at how regulatory delivery systems can seek to consistently support (and use) cultural change across an entire sector or field, looking at the experience of food safety.

Despite the fact that regulators and food businesses are aware of the existence of food hazards and have (respectively) imposed and adopted many preventive practices, the threat of food borne diseases continues to be a source of concern. This need not, in fact, mean that food safety is quantitatively speaking a major risk at present in developed countries or OECD members, though it still definitely is thus in a number of low-income countries. One of the difficulties in discussing the importance of food safety risks is that risk perception is not necessarily correlated to actual, quantifiable risk, as has been evidenced in a number of studies and cases (Slovic, 1986[6]). Another challenge is that assessing the actual prevalence of food borne diseases is far from easy, even in countries with advanced health systems and robust food safety regulation (Blanc, 2021[7]). This being acknowledged, and in spite of food being predominantly safe in developed economies (and in many transition countries as well), food safety risks remain both objectively significant, and acutely present in public perception.

In the United States, it is estimated that over **48 million cases of foodborne illnesses occur annually, with** 128 000 hospitalisations and 3 000 deaths (U.S. FOOD & DRUG ADMINISTRATION, 2018[8]). EFSA estimates that the number of foodborne zoonotic diseases is over 350 000 annually and that the most common causes are *Campylobacter, Salmonella, Yersinia, E. coli* and *Listeria* (European Food Safety Authority, 2021[9]). *Food is exposed to multiple chemical contaminants from the environment or those emerging during its* production, distribution, packaging, or consumption. The accumulated knowledge on hazards **has been leading to the development of preventive programmes aiming to reduce their occurrence in food. However, the effectiveness of such programmes is very much dependent on the structural design and maintenance of facilities along with the human factor, which is responsible for the application of designed practices** (Insfran-Rivarola et al., 2020[10])**.** Behavioural changes to increase the efficiency of hazard analysis and critical control points (HACCP) and other preventive programmes were investigated by multiple experts and findings indicate the need for a further understanding of predispositions and tendencies on the individuals' activities.

The outbreaks of COVID-19 in certain food industries (particularly meat processing), as well as the role of restaurants in outbreaks among consumers, resulting in clusters of the disease, was largely related to management not implementing regulatory requirements (including for profit motives), but also in some cases to workers' poor understanding of safety rules and their logic. Thus, prevention of contamination is dependent on attitudes and behaviour. Conversely, the experience of transforming culture in relation to food safety rules, which the food industry has undergone in the past two to three decades, can be taken as a starting point to think through how "hygiene culture" could be broadly transformed throughout society in order to provide better protection against infectious diseases such as COVID-19. Indeed, the continued prevalence of the disease, and successive infection "waves" in many countries, have been largely correlated to the difficulty in transforming daily practices at work, in transport, in open spaces etc. – and transforming such behaviour, and changing the prevailing culture, is thus a central challenge (Gray et al., 2020[11]) (Howard, 2021[12]) (Mantzari, Rubin and Marteau, 2020[13]).

Successful examples in the food sector show that the effectiveness of change, either when implementing a standard or responding to "invisible threats" such as bioterrorism or a pandemic, is higher in companies which have a strong food safety culture.

Studies of organisational culture show the importance of values, feelings, ideas and socially shared beliefs within one organisation (Robbins and Judge, 2013[14]) and the effect of management structures on organisations´ culture, either through endorsing or neglecting the recommended practices (Alversson M, 2002[15]). In the food area, the food safety culture concept has been introduced through the revised version of the General Principles of Food Hygiene Standard (CAC/RCP 1-1969, version 2020), which further triggered the revision of the Regulation 852/2004 (2020a) and made food safety culture a subject to monitor by control authorities (Lopp, Goebelbecker and Ruff, 2021[16]). The need for these changes in behaviour was obvious for food businesses involved in international trade, and was endorsed by the Global Food Safety Initiative (GFSI)[1] into benchmarked certification programmes (GFSI, 2018[17]).

How good practices have been implemented by the public and private sector

Rules governing good hygienic and manufacturing practices were published in the first version of the Codex Alimentarius General Principles of Food Hygiene Standard in 1969. HACCP principles became widely applied after the BSE crisis in United Kingdom in the 1980s as well as with the *Escherichia Coli* O157:H7 outbreak in the USA in 1993. Although regulators in many countries reacted promptly, endorsing GMP, GHP and HACCP principles as mandatory and auditing their implementation regularly, the hazards continued to emerge. Control bodies' attention was drawn to the problem of inconsistent implementation of GMPs which threatened the effectiveness of food safety programmes based on HACCP principles. The following step in building a more efficient food safety culture was the introduction of the pre-requisite programmes concept, aimed at increasing the adherence to GMPs through monitoring and verification (Manning, Luning and Wallace, 2019[18]).

Attention of regulators and scientists is permanently focused on the three basic food safety hazards (biological, chemical & physical), which, as per the Campden BRI guide (Robert Gaze, 2015[19]), have been complemented with allergens and with radiological hazards more recently. Messages about hazards and associated risks are usually communicated through regulatory documents. However, to raise awareness among the wider population, other communication means are used. In this sense, the WHO published the "Five Keys to Safer Food", a poster and an accompanying manual, publicising them worldwide to highlight universal food safety problems. These being: keep clean; separate raw and cooked; cook thoroughly; keep food at safe temperatures; and use safe water and raw materials (World Health Organization, 2006[20]). This form of communication, translated into over 40 languages, targeted all, from primary producers to processors, handlers, and consumers across the globe, becoming a powerful tool incentivising changes of habits. Many national food safety agencies recognised its success and started to communicate with businesses and the general public through guides, posters and infographics, raising awareness on good practices and gradually modulating habits and attitudes. The tools for communication of good practices together with the risk-based approach to controls in food safety as well as advices provided by official control bodies helped building a general level of understanding of food hygiene and safety, strengthening trust in national food industries and enforcers. The guides convey recommendations applicable by all (e.g. the Canadian Food Inspection Agency´s Guide to Food Safety or the Food Standards Australia and the New Zealand series of manuals and posters) or recommend good practices in a particular industry (such as the UK Food Standards Agency´s Safer Food Better Business for caterers and retailers).

Private food safety standards (especially ISO 22000 series, FSSC, IFS-Food, BRC, but also standards applied in primary production, such as Global GAP) have a significant influence on the level of implementation of food hygiene and safety practices in food business operators. The establishment and work of the GFSI increased comparability of third-party food safety management system (FSMS) audit results, while, at the same time, allowing businesses to choose the standards based on market demands. The perspective, as exposed by GFSI (GFSI, 2018[17]), is to apply a holistic approach to all food safety hazards through the assessment of a range of external and internal factors that influence the effectiveness of Food Safety Management Systems (FSMS) in companies. The number of food safety incidents in companies which had implemented the BRC standards represented a catalyst for the standard setting body to introduce food safety culture in their Global Standard for Food Safety Issue 8,[2] based on the GFSI recommendations, and to start using it in auditing from 2019.

The *Codex Alimentarius* response to a global demand for determinants of food safety culture was the revision of the General Principles of Food Hygiene Standard (CAC/RCP 1-1969, version 2020) and expansion of the list of general principles with a demand for management commitment to food safety. Without management commitment, the principles governing food safety (science-based and preventive approach, hazard analysis, implementation of pre-requisite programmes, control measures on CCPs, scientific validation of control measures, monitoring, corrective measures, verification, documentation, and communication along the food chain) are often inconsistently applied due to managers' prioritisation of other business needs. By proclaiming that: "Fundamental to the successful functioning of any food hygiene system is the establishment and maintenance of a positive food safety culture acknowledging the importance of human behaviour in providing safe and suitable food", the global Codex standard opened the channel for regulators to include food safety culture in their rules.

At the EU level, the regulatory framework for food safety has been found to be comprehensive. One of the regulations, Regulation EC 852/2004 embedded the food safety culture requirements adjusted to the nature and the size of the food business and based on the: a) commitment of the management and all employees; b) participative leadership which has a foundation in clear communication about responsibilities within the organisation; c) continuous learning; d) unobstructed communication within the organisation; e) continuous improvement of the FSMS; e) on time and evidence based planning of resources for food safety implementation and f) compliance with regulatory requirements.

The US FDA New Era of Smarter Food Safety initiative to deliver the Food Modernization Act is based on a) technology enabled traceability; b) smarter tools and approaches to combat outbreaks; c) new business models and retail modernisation and d) food safety culture (FDA, 2020[21]). FDA plans to promote food safety culture within the agency, at regulated subjects and to the public (see Figure 2.1).

Figure 2.1. A modern food safety vision

Source: https://www.fda.gov/media/139868/download.

Food Standards Australia and New Zealand implements activities to promote food safety culture through a set of tools which aid in diagnosing the level of culture and consists of: i) a questionnaire to assess the culture level, ii) a check list to guide the food businesses when implementing food safety culture and iii) a "culture maturity matrix" to be used for self-assessment of strengths and weaknesses and the progress over time.[3] In the province of Victoria, Dairy RegTech, the dairy regulatory body, implements the analytics of data and of culture to monitor food safety compliance, while the Northern Territory Health, an enforcer, helps businesses improve compliance by providing advice on how to strengthen business culture (Food Standards Australia New Zealand, 2017[22]).

Food safety culture as a part of corporate culture

Changes in corporate culture are usually influenced by some strong external or internal force and the need to secure competitive advantage. They require clear vision, personal involvement of managers and sometimes changes in their own beliefs. This is necessary to be able to secure resources on time and in appropriate quantities, whilst at the same time, motivating employees to be part of the change. Transformational changes, closely connected with learning and innovation, are a long-term process which lasts if the change is needed and has a built-in component of constant adaptation of the company to the external and internal environmental demands (Cummings and Worley, 2009[23]).

External demands, such as regulatory, put a certain degree of pressure on businesses to obey rules. Behavioural studies have shown that people respect rules when they are aligned with their moral values and when rules are enacted and implemented fairly (Hodges C., 2016[24]). Only when the opinions and needs of businesses and the public are addressed in those, such requirements are considered as fair. Relative to the implementation of rules, a consistent and proportionality-based approach on risk ensures fairness. Trust in regulators is achieved when both businesses and regulators share the same ethical values (OECD, 2014[25]).

In corporate settings, internal factors drive the implementation of regulations and their translation into business practices. Food production and handling is burdened with numerous and strict external rules and standards leaving limited space for individuals to show their abilities and creativity. Corporate culture usually found in food businesses is one which favours the top-down approach, thus not empowering workers to participate in development of practices and therefore merely doing what managers have asked for. A critical element is to increase food workers' understanding of requirements depends and how to apply them, and the implementation of private food safety management standards can a huge step towards increasing the participation of workers in the translation of rules and standards into actual work procedures and practices. Will the implementation of rules and standards be effective depends on how the management and workers co-operate, being the management and the standard's development team aware of habits and cultural norms and what is the business' set of core values. The success of a business depends on the trust of external and internal stakeholders and increasingly became an important competitive advantage.

Box 2.1. Food Standards Australia and New Zealand: "ethical business" concept

The ethical business concept15 identifies evidence of trust which businesses shall provide:

- ethical principles at the company level, mandatory for all business units and applied continuously regardless of any management changes;
- consistent implementation of rules and standards as proven by their audits;
- history of no penalties;

- numerous loyal and satisfied customers;
- obtaining regular feedback from customers and staff;
- having structures enabling decisions to be debated to test ethical compliance, evaluated against external views, and made transparent.

The more evidence can the business provide, the higher is the level of trust.

Many global food companies introduced food safety culture as a part of their corporate culture, and thus defined their attachment to the ethical business concept driven by external demands (e.g., supply chain, socio-political, legal and national factors) or by internal context factors (e.g., product, production, and organisational characteristics).[7] They have developed their food safety culture even before the Codex standard or private standards embedded it. Danone corporate governance method included food safety culture in their corporate governance method already in 2014 in order to secure uniform implementation of its food safety policy across all facilities (Frédéric, 2017[26]). Nestlé Corporate Business Principles, mandatory for all Nestlé employees, are supported by the company´s Code of Business Conduct and other policies and integrated in business planning, activities, operations, performance reviews and auditing (Nestlé - Group Legal and Compliance, 2020[27]). New Zealand dairy producer Fonterra introduced food safety culture after the outbreak of botulism, caused by their products in 2013, to safeguard the brand name and reinforce its FSMS. Differently, the initiative of Cargill was driven by the high turnover level of employees which created the need to train them in food safety in a way to secure the continuum of practices and employees' understanding (Scattergood, 2018[28]).

RASFF, TRACES and other platforms as drivers of food safety culture

Every year, numerous RASFF alerts are issued about microbiological, chemical (including allergens) and physical contaminants. In 2019, there were 1 175 alerts. In 2017, the fipronil contamination incident spread in over 56 countries and resulted in 109 RASFF notifications for eggs and 8 for other products (European Commission, 2018[29]). In 2018, deliberate physical contamination of strawberries (with needles) in Australia raised a serious concern on the capacities of the HACCP system to defend food safety and undermined the trust in products coming from Australia.

The TRACES system is an EU online platform with the purpose of facilitating communication between different competent authorities and detecting food fraud.[4] It enables fast tracing of consignments and identification of non-conformant ones.

The Administrative Assistance and Cooperation system – Food Fraud is an IT platform created after the 2013 horse meat scandal and used by EU Member States, Switzerland, Norway and Iceland to exchange information on non-compliance and potential intentional violations of the EU agri-food chain legislation. The number of requests is constantly rising, going from 157 in 2016 to 178 in 2017, experiencing a sharp increase in 2018 (234 request) and reaching 292 in 2019 (European Commission, 2020[30]). Mislabelling is the most common non-compliance with an incidence rate of 47.3% while 36% of requests are due to unapproved treatment and/or processes and replacement/dilution of components in the products.

Horsemeat scandal in 2013 and Salmonella in Lactalis baby formulas in 2017 and 2018 were detected and communicated through RASFF and TRACES. Both cases showed the lack of food safety culture at the side of businesses and that of regulators. Inconsistent implementation of EU traceability rules and lack of fairness in the horsemeat scandal triggered the change in approach to fraudulent practices in marketing at EU level. During the Lactalis outbreak (Jourdan-da Silva et al., 2018[31]), traceability issues impaired communication through RASFF. Only ten countries notified the contaminated lots in the first instance, while problems with tracing of contaminated lots in another 35 countries caused late notifications in the RASFF and impaired the efficiency of the recall.

Evaluation of food safety culture

In transitional economies, underdeveloped and sometimes outdated food legislation together with weak capacities of enforcement agencies fail to create the supportive external environment for food safety. An investigation into food safety culture elements in Zimbabwe shows the effect of both internal (food safety programme in place, products' risk level and the resilience of the food production system) as well as external environment determinants (national values and food safety governance characteristics) (Nyarugwe, 2020[32]). This research found that inadequate governance and inconsistency in enforcement resulted in reactive food safety business culture, based only on *ex post* (corrective) reactions, no matter the product risk level, particularly in vulnerable food production systems.

An investigation in 470 businesses from ten Central and East European countries (EU Member States and third countries) showed that preventive systems based on certified FSMSs were aligned with higher understanding of food safety. If the level of knowledge of food safety was higher in EU Member States, the attitudes towards hygiene and food safety of managers were the same in MSs and in Third countries, with all managers taking seriously legislative requirements and prioritizing investment in hygiene and food safety over other business needs (Tomasevic et al., 2020[33]).

In the role-based culture, typical of transitional countries, strict hierarchal and bureaucratic approach to food safety is found. It is associated with reactive food safety culture and only *ex post* (corrective) actions. In more task-oriented cultures, with food safety based on the risk paradigm, managers' positions and influence are based on knowledge and experience. The HACCP principles and FSMSs contribution to the task-oriented culture is displayed in the need for all workers to be well trained and informed about hygiene and food safety and in the empowerment of those who are members of the HACCP team or responsible for monitoring, corrective actions and verification. Task-oriented cultures are thus more associated with pro-active and active food safety culture.

With the rising evidence that the elements of culture influence the preventive approach and the effectiveness of food safety programmes, the need for assessment and measurement of food safety culture emerged.

A range of quantitative (data obtained through questionnaires) and qualitative methods (focus groups, interviews, discussion groups) have been used to assess food safety culture in business settings. Often, a combination of the two is used to assess more precisely the culture. The information which can be obtained through the assessment may help a) improve compliance to rules; b) allow comparing the culture between different facilities owned by the same entity and help the management target those where improvements are needed; c) help define training needs; d) increase awareness of food safety; e) support management and employees commitment to food safety and f) identify weaknesses in FSMS and the level of risk for food safety.

Evaluation tools for businesses

The investigation on the *Listeria* outbreak within Maple Leaf Foods Inc. deli products in 2008, when 23 persons died in Canada, identified the insufficient commitment of the management of the company to food safety. This insufficiency caused inappropriate risk assessment. The company failed to recognise and identify the underlying cause of a sporadic but persistent pattern of environmental positive test results for *Listeria spp* since managers had not Listeria as a priority in their HACCP plans. The new management strongly committed to food safety and included sustainable food safety behaviour into the existing company culture. This included a combination of the emphasis on technical conditions for food safety and on behavioural factors which influence managers' attitudes. Several studies (Powell, Jacob and Chapman,

2011[34]) (Jespersen, Griffiths and Wallace, 2017[35]) assessed different food safety culture evaluation models used in food businesses and identified five main dimensions of culture which may provide information about needs for improvements:

- Values and mission – they need to be defined having in mind the long-term effect on food safety with leaders understanding and supporting food safety;
- People – external and internal stakeholders influence and engagement, education, governance, motivation and communication;
- Consistency – of food safety proprieties with people, technology, resources and processes;
- Adaptability – to the ever-changing environment, and
- Risk awareness – including how risks are managed and communicated.

These dimensions clearly show the connection between the corporate and the food safety culture, since only risk awareness is specific for food safety and the other four are common for both cultures. These five dimensions have been adopted i.a. by GFSI and broken down in sub-dimensions to be used to understand better and improve food safety culture, and gradually integrated into leading FSMS standards. The review of 41 published papers on the evaluation of food safety culture demonstrated that the use of more determinants of culture provides a more complete picture (Samuel, Evans and Redmond, n.d.[36]).

Upon defining the culture, GFSI suggests applying a "food safety maturity matrix", a tool based on "attributes" of leadership corresponding to each maturity level (Table 2.1). The maturity is closely connected to leaders' and managers' attitudes towards food safety and hygiene and is reflected in workers' behaviour and compliance with rules.

Table 2.1. GFSI maturity level matrix

Maturity level	Calculative non-compliers	Managers encourage non-compliance except when there is a risk of enforcement. They do not secure resources for hygiene.
	Doubting compliers	Managers themselves do not follow the rules and do not provide feedback to employees when they fail to follow the rules.
	Dependent compliers	Inconsistent leadership regarding food hygiene. Following rules from the regulator. Lack of initiative and presence of leaders/managers in food working areas only during official controls.
	Proactive compliers	Leaders and managers follow rules and provide good example to employees. Leaders provide feedback to employees regarding compliance with legislative requirements.
	Leaders	Active support to employees, frequent encouragement to apply hygiene procedures, recognition of good practices implemented by employees.

Source: Adapted from (GFSI, 2019[37]).

Simplot Australia, a key wholesaler, introduced its own numerical maturity scale to measure food safety and quality culture in terms of: a) awareness; b) roles and responsibilities associated with food safety and quality; c) cross-functional ownership of food safety and quality outcomes and d) decision-making authority of food safety and quality at all operational levels. Once the level is measured, it is compared to targets and used for strategic planning, to complement other measurable indicators such as financial success and customers' satisfaction (Food Standards Australia New Zealand, 2017[22]).

It is important to emphasise that having only food safety management systems is not enough protection from food-related incidents. A strong food safety culture extends the responsibility to all those involved in the business process and allows managers to do their work, instead of "extinguishing fire" caused by employees not performing their food safety-related tasks. International food businesses (processors, restaurants, retailers), which are good performers, openly communicate their food safety practices to customers, third party auditors and regulators and this transparency provides a framework to increase trust of external stakeholders (Manning, 2018[38]).

Evaluation tools for control agencies

An example of an enforcer using the assessment culture when choosing the approach to a business is the UK Food Standards Agency Food Safety Culture Diagnostic Toolkit for Inspectors, which guides inspectors through: a) the assessment practices b) the use of the food safety culture matrix to categorise the culture in a business and c) the provision of advice to businesses on how to improve the culture (Food Standards Agency, 2012[39]). The Toolkit should help inspectors understand the management's attitudes to hygiene and food safety and its relation with the company's compliance. Inspectors may choose to perform a more general or more in-depth assessment of a business based on observations of the food safety elements and behaviour of managers. Behaviour of managers is to be classified into one of the five categories, namely: a) calculative non-compliers; b) doubting compliers; c) dependent compliers; d) proactive compliers and e) leaders. Each category can be further investigated across food safety elements of a business: a) priorities and attitudes towards food safety and hygiene; b) perception and knowledge of food safety hazards; c) confidence in food safety requirements; d) ownership of food safety and hygiene; e) competence, learning and training in food safety and hygiene systems; f) employee engagement in review and development of food hygiene practices and g) communications and trust to engage in food safety and hygiene report issues. The goal of the Toolkit is to: a) increase compliance of businesses by providing them the type of advice which is the most fit for each category of managers ´behaviour; b) provide advice for each food safety element of a business per category of managers´ behaviour; and c) integrate all food safety elements of a business in one matrix.

The value of the FSA Toolkit has been examined in a survey done on a sample of environmental health officers, food and beverage managers and academics. The findings suggest that external assessment helps identify gaps which otherwise would be hidden (mostly in small-scale businesses, where food safety was not a matter of high concern). The survey also highlighted certain attitudes of line managers in food businesses, which may hinder implementation of food safety management systems. Lastly, having the Toolkit, businesses may perform self-audits and correct attitudes on time, thus increasing the efficiency of food safety management systems (Nayak and Waterson, 2017[40]).

Australian dairy regulator RegTech has its own qualitative scheme for assessing maturity. It is based on the Jesperson's five dimensions and classifies maturity in five stages: Doubt, React, Know, Predict, and Internalise. The stages correspond to the GFSI´s levels of maturity and FSA categories and the goal of this scheme is to identify the linkage between behaviours and the five dimensions of culture and to improve food safety by changing behaviours.

Food safety culture and COVID-19

The US FDA New Era of Smarter Food Safety initiative highlighted the need for new technology. COVID-19 revealed the need to use technological solutions which can store and process food safety big data in many countries. There is not always, however, effective sharing of the data collected respectively by official agencies (regulators) and by private businesses and private certifiers. There is potential for potentially improved food safety management and regulation if data were to be shared more regularly and effectively, in particular regarding food safety culture.

COVID-19, by requiring physical distancing, influenced the communication, training and coaching methods and their frequency. Multiple food testing laboratories offered their services to health authorities and thus their capacities for testing food became limited. This induced the FDA's recommendation to businesses to perform only essential food safety verification tests and implement more stringent control on cleaning and disinfection and control of workers´ habits and health (U.S. Food & Drug Administration, 2021[41]). Such new requirements are better addressed in active and adaptable food safety cultures (pro-active compliers and leaders) where the synergism of attitudes and behaviour along with consistent and structured approach to food safety delivers better results.

Improved knowledge of businesses' food safety culture could help regulatory agencies perform more accurate risk-based categorisation of businesses and allow to reduce the frequency of physical inspections, reducing the exposure of inspectors and businesses to the virus.[5]

Conclusions

The evolution of controls in respect of food safety from *ex post* to *ex ante* resulted in a better understanding of hazards and ways to favour their mitigation and elimination. Although HACCP principles are mandatory in many countries and risk-based approach to control is becoming widely spread, the number of food incidents remains high and new hazards emerge. The level of activity and maturity of food safety culture is a significant component of FSMSs and food public agencies' control. Food safety culture is as significant as the implementation of HACCP principles was in due time.

Certification bodies have significantly contributed to the improvement of food safety, but still, food incidents continue to emerge. The dilemma here is not whether to implement food safety culture, but how to do it. The initiative of GFSI to include food safety culture in the standard and to perform the assessment when auditing for certification purposes is a significant step towards having more businesses improving the culture. Having the same elements of culture, identified by control agencies and GFSI, suggests that control agencies may use the maturity level determined by third party auditors to adapt their advice and control activities to certified businesses.

The new era of food safety is food being reformulated: there are new foods, new production methods, and new delivery methods along with an increasingly digitised system.

Notes

[1] The Global Food Safety Initiative is a business-driven initiative for the continuous improvement of food safety management systems to ensure confidence in the delivery of safe food to consumers worldwide. GFSI provides a platform for collaboration between some of the world's leading food safety experts from retailer, manufacturer and food service companies (…) Key activities within GFSI include the definition of food safety requirements for food safety schemes through a benchmarking process. This process is thought to lead to recognition of existing food safety schemes and enhances confidence, acceptance and implementation of third party certification along the entire food supply chain (source: https://en.wikipedia.org/wiki/global_food_safety_initiative).

[2] https://www.brcgs.com/our-standards/food-safety/.

[3] https://www.foodstandards.gov.au/foodsafety/culture/Pages/default.aspx.

[4] TRACES is the European Commission's multilingual online platform for sanitary and phytosanitary certification required for the importation of animals, animal products, food and feed of non-animal origin and plants into the European Union, and the intra-EU trade and EU exports of animals and certain animal products.

⁵ In this context, it is relevant to mention the European Commission Implementing Regulation (EU) 2020/466, passed as temporary measures to contain risks to human, animal and plant health and animal welfare during certain serious disruptions of Member States' control systems due to COVID-19, https://eur-lex.europa.eu/legal-content/en/txt/?uri=celex%3a32020r0466.

References

Alversson M (2002), *Understanding Organizational Culture*, Sage Publications LTD. [15]

Baldwin, R. (1990), "Why Rules Don't Work", *The Modern Law Review*, Vol. 53/3, pp. 321-337, http://dx.doi.org/10.1111/j.1468-2230.1990.tb01815.x. [1]

Cummings, T. and C. Worley (2009), *Organization Development & Change*, South Western Cengage Learning. [23]

European Commission (2020), *2019 Annual Report: The EU Food Fraud Network and the Administrative Assistance and Cooperation System*, European Union, Luxembourg, https://ec.europa.eu/food/sites/food/files/safety/docs/ff_ffn_annual-report_2019.pdf (accessed on 26 February 2021). [30]

European Commission (2018), *2017 Annual Report: The Rapid Alert System for Food and Feed*, European Union, Luxembourg, https://op.europa.eu/en/publication-detail/-/publication/f4adf22f-4f7c-11e9-a8ed-01aa75ed71a1/language-en/format-PDF/source-174743070 (accessed on 26 February 2021). [29]

European Food Safety Authority (2021), *Foodborne Zoonotic Diseases*. [9]

FDA (2020), *New Era of Smarter Food Safety FDA's Blueprint for the Future*, FDA, https://www.fda.gov/media/139868/download (accessed on 9 February 2021). [21]

Food Standards Agency (2012), *Food Safety Culture Diagnostic Toolkit for Inspectors*, Food Standards Agency , London, https://www.food.gov.uk/sites/default/files/media/document/803-1-1431_FS245020_Tool.pdf (accessed on 26 February 2021). [39]

Food Standards Australia New Zealand (2017), *Food Safety Culture Connections*. [22]

Frédéric, R. (2017), *Same Goal, New Path: Managing Food Safety in an International but Locally Focused Company Like Danone*. [26]

GFSI (2019), *A Culture of Food Safety. A Position Paper from the Global Food Safety Initiative (GFSI)*, https://mygfsi.com/wp-content/uploads/2019/09/GFSI-Food-Safety-Culture-Full.pdf. [37]

GFSI (2018), "A Culture of Food Safety: A Position Paper from The Global Food Safety Initiative (GFSI)", https://mygfsi.com/wp-content/uploads/2019/09/GFSI-Food-Safety-Culture-Full.pdf (accessed on 26 February 2021). [17]

Gray, L. et al. (2020), "Wearing one for the team: views and attitudes to face covering in New Zealand/Aotearoa during COVID-19 Alert Level 4 lockdown", *Journal of Primary Health Care*, Vol. 12/3, p. 199, http://dx.doi.org/10.1071/hc20089. [11]

Hart/Beck (ed.) (2018), *Ethical Business Practice and Regulation*. [4]

Hodges C. (2016), "Ethics in Business Practice and Regulation". [24]

Hodges, C. (2015), *Law and Corporate Behaviour*, Hart/Beck. [3]

Howard, M. (2021), "Gender, face mask perceptions, and face mask wearing: Are men being dangerous during the COVID-19 pandemic?", *Personality and Individual Differences*, Vol. 170, p. 110417, http://dx.doi.org/10.1016/j.paid.2020.110417. [12]

Insfran-Rivarola, A. et al. (2020), "A Systematic Review and Meta-Analysis of the Effects of Food Safety and Hygiene Training on Food Handlers", *Foods*, Vol. 9/9, http://dx.doi.org/10.3390/foods9091169. [10]

Jespersen, L., M. Griffiths and C. Wallace (2017), "Comparative Analysis of Existing Food Safety Culture Evaluation Systems", *Food Control*, Vol. 79, http://dx.doi.org/10.1016/j.foodcont.2017.03.037. [35]

Jourdan-da Silva, N. et al. (2018), "Ongoing nationwide outbreak of Salmonella Agona associated with internationally distributed infant milk products, France, December 2017", *Eurosurveillance*, Vol. 23/2, http://dx.doi.org/10.2807/1560-7917.es.2018.23.2.17-00852. [31]

Lopp, S., J. Goebelbecker and P. Ruff (2021), "The draft of the Regulation (EC) No 852/2004: food safety culture under new administration", *Journal of Consumer Protection and Food Safety*, Vol. 16/1, pp. 93-96, http://dx.doi.org/10.1007/s00003-020-01310-0. [16]

Manning, L. (2018), "The Value of Food Safety Culture to the Hospitality Industry", *Worldwide Hospitality and Tourism Themes*, Vol. 10/3, http://dx.doi.org/10.1108/WHATT-02-2018-0008. [38]

Manning, L., P. Luning and C. Wallace (2019), "The Evolution and Cultural Framing of Food Safety Management Systems—Where From and Where Next?", *Comprehensive Reviews in Food Science and Food Safety*, Vol. 18/6, http://dx.doi.org/10.1111/1541-4337.12484. [18]

Mantzari, E., G. Rubin and T. Marteau (2020), "Is risk compensation threatening public health in the covid-19 pandemic?", *BMJ*, p. m2913, http://dx.doi.org/10.1136/bmj.m2913. [13]

Nayak, R. and P. Waterson (2017), "The Assessment of Food Safety Culture: An investigation of Current Challenges, Barriers and Future Opportunities within the Food Industry", *Food Control*, Vol. 73, http://dx.doi.org/10.1016/j.foodcont.2016.10.061. [40]

Nestlé - Group Legal and Compliance (2020), *Corporate Business Principles*. [27]

Nyarugwe, S. (2020), *Influence of Food safety Culture on Food Handler Behaviour and Food Safety Performance of Food Processing Organisations*, Wageningen University, http://dx.doi.org/10.18174/504736. [32]

OECD (2020), *Behavioural Insights and Organisations: Fostering Safety Culture*, OECD Publishing, Paris, https://dx.doi.org/10.1787/e6ef217d-en. [5]

OECD (2014), *The Governance of Regulators*, OECD Best Practice Principles for Regulatory Policy, OECD Publishing, Paris, https://dx.doi.org/10.1787/9789264209015-en. [25]

Powell, D., C. Jacob and B. Chapman (2011), "Enhancing Food Safety Culture to Reduce Rates of Foodborne Ilness", *Food Control*, Vol. 22/6, http://dx.doi.org/10.1016/j.foodcont.2010.12.009. [34]

Robbins, S. and T. Judge (2013), "Organisational Behavior, 15th Edition", *Pearson.* [14]

Robert Gaze (2015), *HACCP: A Practical Guide*, Campden BRI. [19]

Russell, G. (2019), *Regulatory Delivery*, Hart/Beck. [2]

Samuel, E., Evans and E. Redmond (n.d.), *Assessing Food Safety Culture in Food-Manufacturing: A Review of Applicable Determinants and Tools*, https://www.cardiffmet.ac.uk/health/zero2five/research/Documents/ESamuel%20IAFP%20EU%20Assessing%20FSC%20in%20Food%20Manuf%20FINAL.pdf (accessed on 26 February 2021). [36]

Scattergood, G. (2018), *Creating a Food Safety Culture, part 1: Key takeaways from Fonterra and Cargill*, https://www.foodnavigator-asia.com/Article/2018/03/20/Creating-a-food-safety-culture-part-1-Key-takeaways-from-Fonterra-and-Cargill (accessed on 26 February 2021). [28]

Slovic, P. (1986), "Informing and Educating the Public About Risk", *Risk Analysis*, Vol. 6/4, pp. 403-415, http://dx.doi.org/10.1111/j.1539-6924.1986.tb00953.x. [6]

Tomasevic, I. et al. (2020), "Validation of Novel Food Safety Climate Components and Assessment of Their Indicators in Central and Eastern European Food Industry", *Food Control*, Vol. 117, http://dx.doi.org/10.1016/j.foodcont.2020.107357. [33]

U.S. FOOD & DRUG ADMINISTRATION (2018), *What You Need to Know about Foodborne Illnesses*. [8]

U.S. Food & Drug Administration (2021), *Food Safety and the Coronavirus Disease 2019 (COVID-19)*, U.S. Food & Drug Administration. [41]

van Rooij, B. (ed.) (2021), *Using outcomes to measure compliance: Justifications, challenges, and practices*, Cambridge University Press. [7]

World Health Organization (2006), *Five Keys to Safer Food Manual*, World Health Organization. [20]

3 Implementing food safety regulation and third-party certification in crisis situation

This chapter analyses the challenges that the COVID-19 crisis presents for the implementation and enforcement of food safety regulation. It discusses, in particular, different approaches adopted by food safety authorities to safeguard sustained compliance with food safety regulation under unprecedented circumstances including measures aimed at third-party auditors and certification bodies that play a role in ensuring the food security objectives are achieved.

The challenges for implementation and enforcement of food safety regulation facing COVID-19

The COVID-19 crisis has placed governments under extraordinary pressure to swiftly adjust their regulatory delivery practices, promoting safer practices in workplaces in response to the pandemic, while continuing to ensure effective compliance with rules and regulations across policy fields. Interventions were needed to provide guidance to stakeholders and adopt regulatory easement measures to ensure supply of essential goods and the continuous operation of services or alleviate the regulatory burden on market actors. Regulators have also adapted their routine functions, including inspection and enforcement arrangements, to unprecedented circumstances that restricted the movement of people and face-to-face interactions (OECD, 2021[1]).[1]Several regulators across different streams have leveraged on technology to remain flexible, agile and resilient and also to minimise disruptions arising in trade flows while maintaining protection for healthy and safe trade of food (OECD, 2021[1]). Overall, the crisis sheds light on the importance of the implementation phase of the regulatory governance cycle, an area identified by (OECD, 2018[2]) as a critical and often neglected link in regulatory governance, key to ensure the quality and effectiveness of regulatory policy and to meet the goals of regulations.

The crisis impacted all regulatory areas, including food safety regulation, the set of norms and procedures regulating the production, supply and sale of food with the aim to reduce the risk of unsafe food making consumers ill. Even in normal times, effective food safety compliance combines a complex set of factors: good regulation, well-designed enforcement, and, possibly most importantly, competence, knowledge, and understanding of food safety's importance from food business operators (FBOs) (Blanc and Macrae, 2021[3]).

While there is significant diversity, fragmentation and all kind of "national peculiarities" when it comes to food safety regulatory structures (Blanc and Cola, 2017[4]), across the food supply chain, food safety governance rests on the complementary interplay of public legislation and regulation and with private standards (Kotsanopoulos and Arvanitoyannis, 2017[5]). Public authorities (including national governments, international and transnational institutions) are responsible for setting laws and regulations that define minimum food safety or marketing requirements for food operations. Food-regulation enforcement authorities - structured on a national or subnational basis - then ensure that market actors follow these requirements deploying regulatory delivery tools: inspections and enforcement.

To ensure the safety and quality of their products, food business operators (FBOs) have developed a set of private food safety standards (PFSS). Whether or not required under law, these standards are followed by a large share of the food sector, in particular suppliers integrated into large national or international supply chains, being imposed by buyers (large producers/processors, and large retailers) on their suppliers to ensure the quality and safety of products (Fagotto, 2015[6]). Many food companies implement a systematic preventive approach (based on hazard analysis and critical control points [HACCP]) for their food safety management system. These are frequently based on standards developed and managed by the food industry and operate at different levels, for instance the Safety Quality Food (SQF) standard managed by the US-based Food Marketing Institute and the British Retail Consortium Global Standards for Food Safety (BRC) standard by the British Retail Consortium. Internationally, institutions such as the Global Food Safety Initiative (GFSI) benchmark different international food safety certification programmes and auditing platforms aiming to set equivalency among standards. In addition, international standardisation organisations such as the International Standardization Organisation [ISO] also develop standards relevant for food safety (in particular the ISO 22000 standard for food safety management, including both so-called "pre-requisites" i.e. good hygiene practices, and HACCP implementation).

Monitoring and enforcement of PFSS involves a broad range of actors: scheme owners authorise third-party certification bodies that carry out audits on facilities and issue certificates to attest compliance with a standard. Accreditation bodies provide an extra layer of assurance over the impartiality and capabilities of certification bodies to perform their functions (Figure 3.1).

Figure 3.1. Actors involved in monitoring and enforcement of food safety standards

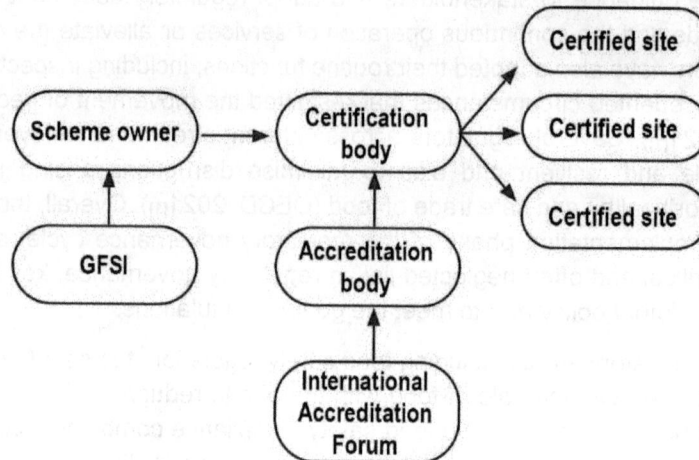

Source: (Fagotto, 2015[6]).

The COVID-19 crisis increased complexity food safety compliance, forcing authorities to adjust their implementation and enforcement practices while also dealing with a set of additional and specific challenges – in particular, strong limitations on the possibility to conduct on-site inspections, changes in distribution and consumption patterns imposed by lockdowns, abrupt changes in both supply and demand, and the need to ensure the food industry's continued functioning in spite of a major economic slump.

Facing food supply chains disruptions

The outbreak of the COVID-19 pandemic affected food supply and demand in complex ways, creating bottlenecks across the food supply chains, impacting farm labour, processing, transport and logistics (OECD, 2020[7]) (OECD, 2020[8]). As illustrated throughout the paper, the SDGs provide an effective framework capable of highlighting the broader implications of changing food supply and demand. In this regard safety enabler function in achieving the goal SDG 5, 8, 11, 12, 14, 15. Yet, the advancement of the SDGs might be weakened by food safety itself, for instance when regulations in this field act as non/tariff sures, having consequences on food exporters in developing countries who have limited capacity and resources to comply with them. The *OECD Agricultural Policy Monitoring and Evaluation 2020* discusses the wide set of agriculture and food policy responses introduced by governments in response to the virus and associated mitigation measures that included mobility restrictions (Box 3.1). These responses were varied and included the provision of various forms of support to actors along the food chain, initiatives to keep food and agricultural supply chains moving, and support to consumers and vulnerable populations, among other. Several countries took active steps to facilitate trade, although some countries also introduced export restrictions in efforts to ensure availability on domestic markets.

Box 3.1. Agriculture and food related measures taken by governments in response to the COVID-19 crisis

The *OECD Agricultural Policy Monitoring and Evaluation 2020* identifies a diverse set of agriculture and food related measures that have been taken by governments in response to the crisis, focusing on agricultural production, the functioning of the food chain and consumer demand. A review of the more than 400 collected policy responses suggests seven broad categories of measures:

1. Sector-wide and institutional measures – including declaration of essential sector and measures related to the functioning of the government.
2. Information and co-ordination measures – such as websites and campaigns, monitoring the agriculture market, co-ordination with the private sector and international co-ordination.
3. Measures on trade and product flows – including trade easing measures, logistics and transport facilitation measures, trade restricting measures, rechannelling product flows, and facilitating internal market integration.
4. Labour measures – notably measures to ensure the health of workers and agriculture labour measures.
5. Agriculture and food support measures – such as general financial sectoral support, specific product support and administrative and regulatory flexibility.
6. General support applicable to agriculture and food – including those provided by overall economic measures and social safety nets.
7. Food assistance and consumer support – including food assistance and market measures in support of consumers.

Source: (OECD, 2020[9]).

Responding to changes in consumer preferences

The downfall in consumption of food away from home triggered shifts in consumer demand with food supplied to supermarkets in lieu of restaurants and other food service establishments. A survey of the impact of COVID-19 in consumer food behaviours in ten EU countries, showed an overall 45% increase in online shopping compared to the pre-COVID-19 period with particular rises in food delivery and bulk purchases (EIT Food, 2020[10]). The EIT data also showed changes in consumer product preferences from the loss of income, an increase in preference for fruit, vegetables and flour, followed by dairy products and poultry, a shift towards pre-packaged food and more attention to date of packaging/best before/expiry date freshness of products for freshness and avoidance of artificial flavours and additives.

The increase in online grocery shopping and food delivery impacted packaging and labelling requirements overseen by food regulation authorities while at the same time calling for special supervision of this trade channel to ensure control food safety and prevent fraudulent practices. The pivot to online food shopping and delivery has increased pressure over food regulators to ensure that new food sellers entering the market meet food safety requirements.[2]

Minimising contamination risks among food sector workforce

Food businesses faced particular challenges to control COVID-19. As many manufacturing facilities, locations such as food processing and packing plants, where individuals work in close proximity and physical distance is hard to observe, face particular obstacles in control of infectious diseases. Several countries saw infection clusters in meat and poultry processing and packaging plants leading at times to a temporary closure of facilities or high absenteeism. Data from the US Centers from Disease Control and Prevention showed that between April and May of 2020, 16 233 workers in meat and poultry processing facilities were infected with COVID-19 (Waltenburg et al., 2020[11]).

Food processing plants outside the meat and poultry business were also affected by virus outbreaks. For instance, in February 2021 over 200 workers in an ice-cream factory in Germany tested positive for COVID-19 (The Germany Eye, 2021[12]). The fact that COVID-19 outbreaks among these workers can rapidly affect large numbers of persons increased pressure for targeted guidance and regulatory intervention to reduce the risk for COVID-19 spreads in food plants.

Reacting to communication needs

Since the early days of the pandemic, food safety authorities responded to an increasing number of queries and questions from a range of actors, including other authorities, the food industry, consumers, and stakeholders (FAO/WHO, 2020[13]). Initially, communication efforts were largely aimed at mitigating concerns about COVID-19 being a food borne disease – the European Food Safety Authority (EFSA) and the US FDA were among the regulators issuing notes on this regard. More often, communication efforts have centred on guiding businesses and consumers around the regulatory easement measures adopted to facilitate the operations of food business and industries.

Reducing food safety incidents to ease pressure in health systems

The COVID-19 outbreak subjected health systems to an overwhelming and sudden surge in the number of patients in need of urgent treatment. This made the need for preventing foodborne illness particularly acute to avoid burdening an already stressed emergency medical system during a global health crisis. Yet continuous implementation of the SDGs (as opposed to them being side-lined by the presence of COVID-19 and potentially regress the progress until today) can support sustained recovery. This is particularly true when referring to SDGs such as SDG 3 "Good health and Well-being" which calls for research and development of vaccines and strengthening capacity for early warning and risk reduction (OECD, 2020[14]). It is too early to assess the impact of COVID-19 over the incidence of foodborne contamination: sudden disruptions in supply chains and workforce availability could have led to increased risks and contamination, but improved hygiene could have been a countervailing force, and data points in conflicting directions. The International Food Safety Authorities Network (INFOSAN) has recorded engaged in 127 events in 2020 compared to 84 in 2019, but some national data suggests decreased incidence for at least the first part of 2020.[3] Variations from previous years could be observed from the increased health safety measures for COVID-19 that offer protection for foodborne diseases, such as handwashing, while at the same time under-reporting could occur from a decrease in doctor consultations and potentially less reporting from laboratories.

Adapting operational arrangements

(WHO and FAO, 2020[15]) identified additional operation challenges facing food authorities, as they implemented business continuity plans to ensure continued enforcement of and compliance with food regulations at times dealing with reduced staff capacity to maintain fully functioning inspection activities, as personnel were assigned to national COVID-19 emergency response teams, staff working from home, and staff illness and self-isolation. A similar challenge was noted for food laboratories with food testing capacities reduced as staff were reallocated to COVID-19 clinical testing.

The next section of this chapter discusses the different measures implemented by food safety authorities and stakeholders during the COVID-19 crisis to safeguard sustained compliance with food safety regulation across the broad range of actors that play a role in ensuring the food security objectives are achieved.

Rebooting enforcement of food safety regulation

Administrations responded to the COVID-19 crisis using a range of regulatory instruments, including primary and secondary legislation, as well as adopting non-legislative changes to implement urgent reforms impacting a range of policy areas such as public health, emergency response systems, competition legislation (OECD, 2020[16]). The responses seek, for instance, to ensure that supply chains continue to provide urgently needed goods or to secure the continuity of services. Overall, a number of these measures were implemented following a risk-based approach that prioritised preserving inspection and enforcement efforts in areas where critical risks were the greatest or where inspections were required based on complaints.

This section discusses the different approaches available to food safety authorities to safeguard sustained compliance with food safety regulation during the crisis. Some of these approaches can be studied to propose lessons and possible good practices in the field of food security going forward.

Regulatory easing measures

Many countries swiftly instituted temporary administrative and regulatory flexibilities as part of their crisis response aiming to ease the operations of business and industries. This involved resorting to a range of regulatory instruments, primary and secondary legislation, at times passed using emergency legislation or fast-tracked procedures (OECD, 2020[16]), as well as non-legislative action. Food safety regulation was part of this trend.

In some countries, regulators introduced adjustment to labelling requirements to help keep at bay the impact of supply chain disruptions on product availability. In the United States, the Department of Health and Human Services Food and Drug Administration (FDA) issued a set of guidance allowing for temporary labelling flexibilities under certain circumstances. The agency allowed manufacturers to make minor formulation changes without reflecting them on the package label (U.S. Food and Drug Administration, 2020[17]). To meet increased demand for eggs, the FDA issued temporary flexibility guidance on certain packaging and labelling requirements for eggs sold in retail food establishments (U.S. Department of Health and Human Services Food and Drug Administration, 2020[18]). Additional guidance on labelling was issued to allow restaurants to sell packaged food to consumers directly, or to other businesses for sale to consumers (U.S. Food and Drug Administration, 2020[19]). Similar guidance was passed by the US Department of Agriculture allowing for labelling flexibilities to the Country of Origin Labeling (COOL) requirements to enable the redistribution of food products intended for food service to be sold directly to consumers (U.S. Department of Agriculture, 2020[20]). The Canadian Food Inspection Agency (CFIA) also provided flexibility for certain labelling requirements for foodservice packaged products deemed to have no impact on food safety, as part of a broader temporary suspension of some low-risk suspended activities (Canadian Food Inspection Agency, 2020[21]). Denmark temporarily waived labelling requirements of country of origin and accepted the retail sale of pre-packaged foods without labelling in Danish provided that it complied with the requirements of the Food Information Regulation (Danish Veterinary and Food Administration, 2020[22]).

In-person inspections or other compliance activities were at times scaled-back or halted. Some countries chose to restrict controls to high-risk situations focusing exclusively on critical safety issues, so as to minimise possible virus exposure for inspectors and workers. The CFIA's business continuity plan activated in March 2020, prioritised critical activities while temporarily suspending low-risk activities, such as food inspections and investigations not related to food safety, surveillance or sampling activities (food, plant and animal), inspections of preventive control plans and plant and animal inspections in areas of low risk and labelling and domestic facility inspections (Canadian Food Inspection Agency, 2020[23]). In March 2020, the US FDA postponed most foreign facility inspections and all domestic routine surveillance facility inspections while maintaining only all mission-critical assignments (for instance, domestic for-cause inspection) (U.S. Food and Drug Administration, 2020[24]).

Food safety authorities adjusted the requirements applicable to third parties responsible for evaluating compliance with food regulation through certification or audit mechanisms. Some countries provided flexibility on regulations requiring accreditation bodies and third-party certification bodies to perform certain onsite observations and examinations that were ill advised due to health concerns or unviable due to travel restrictions. The US FDA issued a temporary policy in this regard applicable to participants in the US *Accredited Third-Party Certification Program* under which accredited third-party certification bodies conduct food safety audits and issue food or facility certifications to eligible foreign entities for specific purposes (U.S. Food and Drug Administrator, 2020[25]). Regulators also extended certain certificates based on risk assessment of products or facilities. The CFIA allowed for flexibility in Certification Bodies inspection frequency for renewal certification under the Canada Organic Regime. Poland extended the validity of health certificates for livestock and the deadlines for livestock identification (OECD, 2020[9]). Private actors involved in the implementation of food safety regulation took similar flexibility measures. For instance, the GFSI allowed for certificate extension of six month under specific circumstances (GFSI, 2020[26]).

Governments introduced measures to minimise COVID-19 contamination risks safeguarding agriculture and food workforce and to ensure the availability of seasonal labour. Workplace safety is typically overseen by occupational safety and health authorities. Still in a number of countries food safety regulators joined cross-government efforts to provide guidance on special workplace safety and health practices. In the United Kingdom, special guidance for all workplaces involved in the manufacturing, processing, warehousing, picking, packaging, retailing and service of food, aimed to reduce the risk of COVID-19 entering the workplace, spreading within and from the workplace to the wider community and also aimed to reduce the impact of the virus on output and production of the food industry (Public Health England and the UK Department of Environment Food and Rural Affairs, 2020[27]). The UK Food Standards Authority (FSA) provided special guidance to help stakeholders understand how to work safely in the food manufacturing and agricultural sectors during the pandemic (UK Food Standards Agency, 2020[28]) (UK Food Standards Agency, 2020[29]). In the United States, the FDA and the Occupational Safety and Health Administration (OSHA) developed a checklist for human and animal food operations regulated by the FDA to use when assessing operations during the public health emergency, especially when re-starting operations after a shut down or changes in the status of the pandemic (FDA and OSHA, 2020[30]). The checklist is aimed at persons growing, harvesting, packing, manufacturing, processing, or holding human and animal food regulated by FDA, including: produce, seafood, milk, eggs, grains, game meat, and other raw materials or ingredients, as well as their resulting human or animal food products. The document also served to provide information for foreign facilities that manufacture, process, pack, or hold food for consumption in the US. A section of the checklist is focused on food safety, including on food safety or HACCP plans, personnel, suppliers and incoming Ingredients, and current good manufacturing practices (CGMPs) requirements. Additional special guidance issued by the corresponding US authorities in consultation with the FDS centred on meat and poultry and seafood processing workers and employers (CDC and OSHA, 2021[31]) and (CDC and OSHA, 2021[32]).

Deployment of new technologies to enforce food safety regulations

Recent years have seen a growing interest in the potential of new digital technologies, enabled by the increasing use of mobile tools, Machine Learning and Big Data, to increase regulatory capacity particularly in the area of inspections and enforcement (OECD, 2021[1]). The *OECD Regulatory Enforcement and Inspections Toolkit* includes a criteria on information integration notes that Information Communication Technology (ICT) tools should be used to make work more efficient and targeting better (OECD, 2018[33]). The toolkit encourages the use of automated planning (based on risk criteria and risk profile of establishments recorded in the database), mobile tools for inspectors (laptop/tablet or smartphone based tools including check-lists, mobile applications or other instruments to directly record the inspection findings, look up additional information etc.), and geographical information systems (Box 3.2).

Box 3.2. Using technology to support risk-based and outcomes-focused regulatory delivery

Outcomes-focused checklists within the "Rating Audit Control" (RAC) project in Italy

The RAC project in Italy, funded by the European Commission and implemented by the OECD, aims at supporting regional and national governments in improving the business environment and investment climate and the efficiency of the use of public funds through improved regulatory predictability and confidence, and reduced burden on lower-risk activities. To achieve better outcomes in regulatory delivery, inspection methods and practices on the ground are being transformed, consistency of inspections improved, and efforts towards clearer and more understandable regulatory requirements for business operators undertaken.

One key tool to achieve this goal is work on risk-based checklists for inspections, which are being prepared in different regulatory areas and in particular in the food safety domain so as to ensure development and consistency in methods, and to make a valuable contribution to improving matters in terms of outcomes. New checklists are being adapted to regional realities. They include a risk-based scoring system, and their results are being linked to an update in risk rating. By including the "static" risk of the establishment, its "dynamic" risks (actual risk management, such as the use of HACCP in food safety), and its compliance history (including measures imposed by inspectors because of violations leading to immediate risks), they yield a comprehensive picture of the establishment in terms of actual level of risk, and of most significant elements that need to be addressed to achieve the desired outcomes in terms of regulatory goals.

Machine learning and risk indicators

While defining risk abstractly is relatively straightforward, developing robust methods to predict the level of risk of different businesses or establishments is far more difficult. Until recently, challenges in data availability and methods for analysis meant that defining risk criteria and their relative weights based on "data mining" or similar mathematical approaches was mostly reserved to tax and customs inspections (where the objects of regulation and control are inherently numerical, and computerisation was done earliest and in the most systematic way) – in technical, safety and similar fields, risk identification and weighting was done through a combination of scientific and technical findings, regulators' experience and "trial-and-error", but in a much less systematic and precise way.

The spread of information management systems to record inspection results, and thus the increasing availability of detailed historical data, combined with advances in data processing power and analytical tools (e.g. machine learning) now make it increasingly possible. In Italy, the region of Campania is currently piloting the use of Machine Learning for risk assessment in food safety controls.

In addition to such work to better assess "operational-level" risk, work at the "strategic level" is also increasingly data-driven. In 2017, Canada's CFIA launched a review of its risk management model in order to ensure the allocation of resources where it can have the greatest impact on reducing risks. The first challenge of the model is to enable comparison among different kinds of risks, which entails converting different types of risks into comparable data. Based on this, the Agency is able to consider trade-offs among all of them, across different organisational levels This work has entailed considerable efforts to gather and consolidate data from all parts of CFIA's work.

Along similar lines, the Risk Assessment Directorate of Environment and Climate Change Canada has developed the Threat-Risk Assessment (TRA) model, based on a large review of available data to estimate the probabilities and potential impact of known sources of harms for the environment. Data is gathered from the industry, government partners and international actors. Outcomes from the strategic risk assessment are used by the Climate Change and Environment of Canada for project planning and allocation of resources. Likewise, it is shared with enforcement officers to inform their work.

Sharing and using data to better manage risks

A number of Italian regions and institutions have, in recent years, worked on improving data sharing, analysis, and usage, to reduce the burdens and inefficiencies created by duplications and lack of coordination between different services, and better support regional economies.

In Campania, in addition to the existing GISA system to plan and manage all food safety inspections, the region partnered with the University of Naples Parthenope to develop MytiluSE, a system to predict the quality of waters so as to secure safety of mussels produced in the bay of Naples. Rather than expending large resources on *ex post* controls to find potential contamination, the system works pre-emptively, enabling to know which days the harvesting of mussels would be unsafe. Once fully operational, it can both inform producers and guide inspectors' work. Developing the system involved investigating the currents of the bay of Naples, mapping contamination sources, and developing a reliable predictive model, but it is potentially completely transformative for regulatory delivery. It was also adapted to predict air pollution by fumes, which can affect feed for bovine herds. The predictive approach for mussels is not only better for the economy and public service efficiency, but it also avoids health hazards far more effectively, because microbiological testing and sampling takes time, and results can come too late (leading to potential contaminations from other products harvested the same day).

Notes: https://documents.worldbank.org/en/publication/documents-reports/documentdetail/490491468159916971/risk-based-tax-audits-approaches-and-country-experiences; https://www.oecd.org/tax/administration/33818656.pdf%20https://www.oecd.org/tax/forum-on-tax-administration/publications-and-products/hnwi/42490764.pdf.
Source: OECD work in Italy (publications forthcoming 2021), direct interviews with and presentations from CFIA and Environment and Climate Change Canada and Montella R, Riccio A, di Luccio D, Mellone G, de Vita, C G (2020), MytiluSE: Modelling mytilus farming System with Enhanced web technologies, Università degli Studi di Napoli Parthenope, Sciences and Technology Department, commissioned by Campania.

The pandemic increased experimentation with new technologies for regulatory inspections and enforcement, as they provide tools that can help cut through some of the obstacles posed by mitigation measures such as widespread restrictions on travel and mobility, workplace social distancing rules and temporary closure of business. As surveillance of compliance with food safety regulations remained crucial, a number of authorities adapted their "intervention toolkit" following a risk-based approach to ensure compliance using a range of mechanisms, including remote inspections, "record audits" and hybrid inspections, to enable the continuity of verification activities. However, this leap has been unequal, with a significant gap in the usage of new technologies between advanced economies, which could adapt faster than middle- and low-income economies.

Some **authorities responsible for the enforcement of food safety regulations recalibrated their approaches to inspections and relied on remote tools to ensure surveillance continuity.** Key enforcement activities such as on-site inspections were particularly impacted by the outbreak of COVID-19, facing difficulties and restrictions for travelling between and within the borders of countries, accessing food processing and packaging facilities, as well as the need to safeguard inspectors and firm's workforce from health hazards and observing authorities' recommendations. To sort some of these obstacles a number of authorities relied on remote inspections, where visits or checks are conducted by an authorised official in real time using technology without travelling to the site (Box 3.3).[4] Inspections may also take a hybrid form, combining activities that take place remotely, such as interviews and record review, with on-site activities, or setting up inspections that take place remotely and on-site on a later date. This has also been handled as an emergency measure by the EC in its Implementing Regulation (EU) 2020/466.[5]

Box 3.3. Virtual inspections experiments

With new urgency because of the constraints on movement and the need to minimize contagion risks in the COVID-19 context, virtual audits and inspections are under consideration or discussion in a number of countries and institutions, or being piloted to test their reliability and applicability. This is particularly important in food safety, because food production and supply are essential activities that cannot be suspended fully, and food safety inspections are both important to prevent food contamination, but also sources of potential risks of contagion. Challenges in doing such checks remotely include the difficulty to spot "hidden" problems, assurance against fraud, requirements for equipment, training and competence of staff in the facility to provide a "remote view", and of inspecting staff to analyse and challenge the results, etc. In addition, other risks (such as occupational safety and health) cannot be easily observed from outside, and even though they may not strictly belong to "food safety", they nonetheless can have a major negative impact if left unchecked. Such remote audits or inspections are being discussed or trialled in Canada and Italy. On the side of private certification, the GFSI has decided to allow the use of remote audits in certain specific cases and situations (https://mygfsi.com/blog/gfsi-remote-auditing-benchmarking-requirements-updates/): both auditor and audited must agree to use it, the technical conditions must be met, and remote auditing can only apply to a part of the audit but not the entirety of it (thus, it reduces contact while not eliminating it).

Source: OECD, Digitizing Regulatory Delivery using Emerging Technologies – A Review of Current Practices, 2021 (forthcoming) – direct interviews with regulators in Italy and Canada – GFSI website.

In Canada, the CFIA was given special funding to hire, train and equip specialized staff to conduct critical inspections and to develop flexible ways to carry out enforcement activities, notably via the use of digital tools (CFIA, 2020[34]). In April 2020, the US FDA shifted to remote inspections for its Foreign Supplier Verification Programs (FSVP) for Importers of Food for Humans and Animals, which sets certain risk-based activities that importers need to perform to verify that food entering the US has been produced in compliance with applicable US safety standards (FDA, 2020[35]). The agency shifted to reviewing records on electronic format and conducting a limited number of remote inspections, prioritising the inspections of FSVP importers of food from foreign suppliers whose onsite food facility or farm inspections were postponed due to the health emergency.

The increased use of remote verification tools extended to third party audits and certification. In Canada, where the CFIA uses third party oversight system for implementation of the Canada Organic Regime which includes Conformity Verification Bodies (CVBs) and CFIA accredited Certification Bodies (CBs), the agency postponed all planned audits and developed criteria for remote audits to reduce the need for on-site activity by CVBs. Scheme or standard owners, and benchmarking organisations, followed a similar path. For instance, the GFSI amended its benchmarking requirements to allow for the use of ICT as part of the GFSI benchmarked certification audits. The benchmarking document from GFSI provides that the full audit objectives be met and all parts of the audit process completed effectively and as a combined process, while it allows the certification program owners to define the parts of the audit that may be carried out remotely, it requires certain minimum content for the on-site portion, and sets 30 day window for the completion of a full audit, extendable for a maximum of 90 days (GFSI, 2020[26]).

Considerations for the way forward

The response of authorities responsible for enforcement of food safety regulation to the COVID-19 crisis highlights certain considerations for regulatory delivery in this field going forward:

- Regulatory easing measures put in place during the crisis are unlikely to have undergone *ex ante* regulatory impact assessment or stakeholder consultation, still authorities should not forgo the use of good regulatory management tools to ensure scrutiny of these measures' impacts. *Ex post* review should be emphasised to assess whether these measures are delivering on their goals and should be kept in the aftermath of the crisis (OECD, 2020[16]). The *OECD Best Practice Principles of Reviewing the Stock of Regulation* (OECD, 2020[36]) emphasise the importance of relying on clear objectives, data and monitoring processes to assess the impact of regulations. This could form the basis for a review of these emergency easing measures which, if found to have an overall positive impact on economic activity and create no or only negligible additional risks, could be then converted into permanent regulatory changes (or, conversely, reversed if they are found to have created major harms or risks). Additionally, as seen in many countries, the broadening of the scope of traditional regulatory impact assessments can contribute to a coherent SDG implementation, via, for instance, the introduction of Sustainability Impact Assessments (e.g., SIA).

- Pivoting to hybrid and remote inspections presents challenges. Food safety regulators or the corresponding standards owners need to authorise the changes, whether temporarily or permanently, select the supporting technology and provide appropriate guidance and information to inspectors and firms. In addition, inspectors and auditors need to be properly trained and managed to ensure that they have the relevant technical skills to carry out remote inspections.

- Once travel and mobility restrictions are eased or lifted and on-site inspections are authorised to resume, regulators in charge of enforcement and certifiers will need to prepare for an increase in activity and set a risk-based approach to set priorities for inspections. Staff may need to be specially allocated to deal with the backlog of inspections in high-risk establishments.

- Lessons from the adjustments to face the challenges posed by COVID-19 will feed into the approach of inspection and certification bodies to the use of new technologies for food safety oversight. For instance, one of the core elements of the new FDA Blueprint for the Future notes that a smarter approach to inspections includes "Conduct proof-of-concepts to evaluate the feasibility of using remote, virtual, and/or component inspections of foreign and domestic firms with a demonstrated history of compliance for agency prioritization purposes" (FDA, 2020[37]). The blueprint notes that experience of remote inspections conducted for some importers during the pandemic will feed into the FDA's assessment. It also highlights the need to modernise inspection and reporting processes leveraging on mobile inspection technology and digital reporting tools.

- Overall, some of the practices observed around the implementation of food safety regulation during the pandemic could have positive spillovers to strengthen less developed food safety regulatory frameworks, notably in developing countries where they could help address resources and capacity constraints as well as constraints in human resource expertise.

  o Inspection functions should be co-ordinated and, where needed, consolidated to ensure a better use of public resources, minimise the burden on regulated subjects, and maximise effectiveness. Facing staff and resource constrains, remote or hybrid inspections conducted under adequate frameworks could help increase the efficiency of the system and enhance the allocation of expert resources;

  o Food safety inspectors need to be trained and require substantial technical knowledge. Greater collaboration between inspection and certification bodies, and increased reliance on third-party assessment, can help enhance the skills and tools of inspectors building on the expert competences from third-party certifiers;

  o Greater focus on supporting food safety culture and increased attention on management attitudes as an essential part of food safety inspections;

  o Information sharing is key to ensure an optimal use of enforcement resources. Inspection and certification bodies could leverage resources available to third party bodies, particularly evidence and data, to ensure that their surveillance activities are evidence-based and measurement based.

Notes

[1] In the Sanitary and Phytosanitary Systems (SPS), e-certifications helped minimise the negative effects of social distancing arising from the COVID-19 pandemic. This has also helped reduce costs associated with physically handling certifications.

[2] Source: OECD Secretariat interviews with food safety authorities from OECD Member countries.

[3] See https://foodsafety.asn.au/topic/australias-food-safety-report-card-released-for-the-un-world-food-safety-day-7-june-2020/ and https://www.foodsafetynews.com/2021/02/covid-19-measures-accompany-decline-of-foodborne-infections/.

[4] (OECD, 2018[33]) defines "inspection" as any type of visit or check conducted by authorised officials on products or business premises, activities, documents, etc.

[5] https://eur-lex.europa.eu/legal-content/en/txt/?uri=celex%3a32020r0466.

References

Blanc, F. and G. Cola (2017), "Inspections, risks and circumstances. Historical development, diversity of structures and practices in food safety", in *Studi Parlamentari e di Politica Costituzionale*. [4]

Blanc, F. and D. Macrae (2021), "Food Safety Compliance", in *The Cambridge Handbook of Compliance*. [3]

Canadian Food Inspection Agency (2020), *Information regarding certain labelling requirements for foodservice products during the COVID-19 pandemic*, https://www.inspection.gc.ca/covid-19/cfia-information-for-industry/foodservice-products-during-the-covid-19-pandemic/eng/1587075946413/1587075946772 (accessed on 28 January 2021). [21]

Canadian Food Inspection Agency (2020), *The CFIA is prioritizing critical activities during COVID-19 pandemic*, https://www.inspection.gc.ca/covid-19/cfia-information-for-industry/critical-activities-during-covid-19-pandemic/eng/1587076768319/1587076768647. [23]

CDC and OSHA (2021), *Interim Guidance: Meat and Poultry Processing Workers and Employers*, https://www.cdc.gov/coronavirus/2019-ncov/community/organizations/meat-poultry-processing-workers-employers.html (accessed on 6 February 2021). [32]

CDC and OSHA (2021), *Interim Guidance: Protecting Seafood Processing Workers from COVID-19*, https://www.cdc.gov/coronavirus/2019-ncov/community/guidance-seafood-processing.html (accessed on 6 February 2021). [31]

CFIA (2020), *Government of Canada provides $20 million to safeguard Canada's food supply by supporting critical food inspection services*, https://www.canada.ca/en/food-inspection-agency/news/2020/04/government-of-canada-provides-20million-to-safeguard-canadas-food-supply-by-supporting-critical-food-inspection-services.html (accessed on 28 January 2021). [34]

Danish Veterinary and Food Administration (2020), *Coronavirus and food - retail, supermarkets and manufacturing companies*, https://www.foedevarestyrelsen.dk/Leksikon/Sider/Coronavirus-og-foedevarer-Detail-supermarkeder-produktionsvirksomheder.aspx (accessed on 9 February 2021). [22]

EIT Food (2020), *COVID-19 Impact on Consumer Food Behaviours in Europe*, European Institute of Innovation & Technology (EIT). [10]

Fagotto, E. (2015), "Are we being served? The relationship between public and private food safety regulation", in Havinga, T., D. Casey and F. van Waarden (eds.), *The Changing Landscape of Food Governance*, Edward Elgar Publishing, http://dx.doi.org/10.4337/9781784715410.00022. [6]

FAO/WHO (2020), *COVID-19 and Food Safety: Guidance for competent authorities responsible for national food safety control systems*, http://www.fao.org/3/ca8842en/CA8842EN.pdf. [13]

FDA (2020), *FDA To Temporarily Conduct Remote Importer Inspections Under FSVP Due to COVID-19*, https://www.fda.gov/food/cfsan-constituent-updates/fda-temporarily-conduct-remote-importer-inspections-under-fsvp-due-covid-19 (accessed on 8 February 2021). [35]

FDA (2020), *New Era of Smarter Food Safety FDA's Blueprint for the Future*, FDA, https://www.fda.gov/media/139868/download (accessed on 9 February 2021). [37]

FDA and OSHA (2020), *Employee Health and Food Safety Checklist for Human and Animal Food Operations During the COVID-19 Pandemic*, https://www.fda.gov/media/141141/download. [30]

GFSI (2020), *Update on Temporary Audit Measures During Covid-19 Pandemic*, https://mygfsi.com/news_updates/update-on-temporary-audit-measures-during-covid-19-pandemic (accessed on 4 February 2021). [26]

Kotsanopoulos, K. and I. Arvanitoyannis (2017), "The Role of Auditing, Food Safety, and Food Quality Standards in the Food Industry: A Review", *Comprehensive Reviews in Food Science and Food Safety*, Vol. 16/5, http://dx.doi.org/10.1111/1541-4337.12293. [5]

OECD (2021), "Digital opportunities for Sanitary and Phytosanitary (SPS) Systems and the trade facilitation effects of SPS Electronic Certification", *OECD Food, Agriculture and Fisheries Papers*, No. 152, OECD Publishing, Paris, https://dx.doi.org/10.1787/cbb7d0f6-en. [1]

OECD (2020), *Agricultural Policy Monitoring and Evaluation 2020*, OECD Publishing, Paris, https://dx.doi.org/10.1787/928181a8-en. [9]

OECD (2020), *Building a coherent response for a sustainable post-COVID-19 recovery*, OECD Publishing, Paris, https://dx.doi.org/10.1787/d67eab68-en. [14]

OECD (2020), *COVID-19 and the food and agriculture sector: Issues and policy responses*, OECD Publishing, Paris, https://dx.doi.org/10.1787/a23f764b-en. [8]

OECD (2020), "Food Supply Chains and COVID-19: Impacts and Policy Lessons", *Comparing crises: Great Lockdown versus Great Recession*, http://dx.doi.org/10.4060/ca8833en. [7]

OECD (2020), *Regulatory quality and COVID-19: Managing the risks and supporting the recovery*, OECD Publishing, Paris, https://dx.doi.org/10.1787/3f752e60-en. [16]

OECD (2020), *Reviewing the Stock of Regulation*, OECD Best Practice Principles for Regulatory Policy, OECD Publishing, Paris, https://dx.doi.org/10.1787/1a8f33bc-en. [36]

OECD (2018), *OECD Regulatory Enforcement and Inspections Toolkit*, OECD Publishing, Paris, https://dx.doi.org/10.1787/9789264303959-en. [33]

OECD (2018), *OECD Regulatory Policy Outlook 2018*, OECD, Paris, https://dx.doi.org/10.1787/9789264303072-en. [2]

Public Health England and the UK Department of Environment Food and Rural Affairs (2020), *Guidance for food businesses on coronavirus (COVID-19)*, https://www.gov.uk/government/publications/covid-19-guidance-for-food-businesses/guidance-for-food-businesses-on-coronavirus-covid-19 (accessed on 28 January 2021). [27]

The Germany Eye (2021), *Corona outbreak in ice cream factory*, https://thegermanyeye.com/content/amp/corona-outbreak-in-ice-cream-factory-3989.html?__twitter_impression=true&s=03 (accessed on 2 March 2021). [12]

U.S. Department of Agriculture (2020), *USDA Announces Labeling Flexibilities to Facilitate Distribution of Food to Retail Locations*, https://www.ams.usda.gov/content/usda-announces-labeling-flexibilities-facilitate-distribution-food-retail-locations. [20]

U.S. Department of Health and Human Services Food and Drug Administration (2020), *Temporary Policy Regarding Packaging and Labeling of Shell Eggs Sold by Retail Food Establishments During the COVID-19 Public Health Emergency*, https://www.fda.gov/regulatory-information/search-fda-guidance-documents/temporary-policy-regarding-packaging-and-labeling-shell-eggs-sold-retail-food-establishments-during (accessed on 28 January 2021). [18]

U.S. Food and Drug Administration (2020), *Coronavirus (COVID-19) Update: FDA Focuses on Safety of Regulated Products While Scaling Back Domestic Inspections*, https://www.fda.gov/news-events/press-announcements/coronavirus-covid-19-update-fda-focuses-safety-regulated-products-while-scaling-back-domestic (accessed on 28 January 2021). [24]

U.S. Food and Drug Administration (2020), *Temporary Policy Regarding Certain Food Labeling Requirements During the COVID-19 Public Health Emergency: Minor Formulation Changes and Vending Machines*, https://www.fda.gov/regulatory-information/search-fda-guidance-documents/temporary-policy-regarding-certain-food-labeling-requirements-during-covid-19-public-health (accessed on 28 January 2021). [17]

U.S. Food and Drug Administration (2020), *Temporary Policy Regarding Nutrition Labeling of Certain Packaged Food During the COVID-19 Public Health Emergency*, https://www.fda.gov/regulatory-information/search-fda-guidance-documents/temporary-policy-regarding-nutrition-labeling-certain-packaged-food-during-covid-19-public-health (accessed on 28 January 2021). [19]

U.S. Food and Drug Administrator (2020), *Temporary Policy Regarding Accredited Third-Party Certification Program Onsite Observation and Certificate Duration Requirements During the COVID-19 Public Health Emergency*, https://www.fda.gov/regulatory-information/search-fda-guidance-documents/temporary-policy-regarding-accredited-third-party-certification-program-onsite-observation-and (accessed on 28 January 2021). [25]

UK Food Standards Agency (2020), *Adapting agricultural and primary production operations during COVID-19*, https://www.food.gov.uk/business-guidance/adapting-agricultural-and-primary-production-operations-during-covid-19 (accessed on 28 January 2021).   [29]

UK Food Standards Agency (2020), *Adapting food manufacturing operations during COVID-19*, https://www.food.gov.uk/business-guidance/adapting-food-manufacturing-operations-during-covid-19 (accessed on 28 January 2021).   [28]

Waltenburg, M. et al. (2020), "Update: COVID-19 Among Workers in Meat and Poultry Processing Facilities — United States, April–May 2020", *MMWR Morbidity and Mortality Weekly Report*, Vol. 69/27, http://dx.doi.org/10.15585/mmwr.mm6927e2.   [11]

WHO and FAO (2020), *COVID-19 and Food Safety: Guidance for competent authorities responsible for national food safety control systems*, WHO and FAO, Rome, http://dx.doi.org/10.4060/ca8842en.   [15]

4 Optimising food safety regulatory systems for economic recovery

This chapter discusses the opportunities that the COVID-19 crisis presents to rethink and optimise food safety regulation for the recovery. The pandemic has stressed the importance of reducing administrative barriers but also the need for regulations that effectively foster safe practices. Food supply systems showed resilience due to governments´ rapid implementation of temporary measures. Prioritisation and reduction of the number of physical controls did not lead to a safety crisis, and this highlighted the need for greater optimisation and efficiency of controls, and recognition of results of food safety management systems. Progress in technology and data management can help respond to the need for more co-operation and collaboration among control agencies and improved information exchange to improve efficiency and effectiveness of control measures.

Introduction

Regulation played a role at nearly every stage of facing the global health crisis and, going forward, will be a critical element for social and economic recovery. Across policy fields, the pandemic made the need for trusted, evidence-based, internationally co-ordinated and well-enforced regulation particularly acute (OECD, 2020[1]). Governments adapted their use of regulatory management tools, including regulatory impact assessments, stakeholder engagement and *ex post* evaluation and removed a number of administrative barriers to improve regulatory delivery (OECD, 2020[2]).

As described in the previous sections, the COVID-19 pandemic created various specific challenges to food businesses related in particular to supply chains integrity and workers safety. The pandemic showed the benefits of risk-based control systems which helped regulatory agencies cope with reduced physical inspections. Against this background, authorities responsible for food safety regulation used a diverse range of approaches to combine safeguarding the safety of the food supply and adapting to the major difficulties created by the pandemic, including regulatory easement and enforcement through new tools. Over the past years, countries have made major progress in cutting red tape for citizens and business, putting in place more transparent and better regulations, and ways to deliver them. The health crisis and its social and economic aftermaths present a new opportunity to further rethink and optimise regulatory practices and frameworks, including around food safety.

This chapter discusses briefly the impact of COVID-19 on international food trade, consumer preferences and food safety incidents, assesses challenges faced by control agencies and provides recommendation for simplification of regulatory and management processes. The underlying thread for such efforts should be to make food safety regulatory systems more thoroughly risk-based, and improve transparency, communication and stakeholders engagement so that the problems with consumers' (and businesses') trust can be addressed (see Box 4.1). This should be guided by lessons from research and experience, which have helped understand better the drivers of compliance in food safety, and thus better distinguish regulatory instruments that are effective and efficient, from others which may bring more burden than benefits (Blanc and Macrae, 2021[3]).

Box 4.1. Loss of trust in European consumers

The EIT Food TrustTracker study, conducted in 2020 on 19,800 consumers across 18 European countries to measure trust in the food system, showed that farmers are mostly trusted when it comes to fairness and openness of practices (67% of consumers asked trust them and only 13% do not), followed by retailers (53% trust them vs 20% that do not), while 47% of respondents reported trust in regulatory authorities and 46% in manufacturers (while the mistrust expressed was of 29% vs 26%). In relation to the safety of food, 55% of consumers asked consider food as generally safe and 22% as not safe, with over 40% of customers in Turkey, the Czech Republic and Romania regarding food as generally unsafe.

Source: EIT (2020): Food Trust Report. See https://www.eitfood.eu/media/news-pdf/EIT_Food_Trust_Report_2020.pdf.

Challenges faced by food safety regulators during the COVID-19 crisis

The changes observed in 2020 implied that businesses needed to modify their suppliers' management, health and safety procedures and cleaning programs and, in many cases, had to adapt to online sales. Regulators provided guidance to businesses on how to update food safety programs to accommodate for COVID-19-induced changes. The UK Food Standards Agency published an information package to support food businesses with COVID-19 challenges. This included the *Reopening checklist for food businesses during COVID-19*,[1] as well as updated guidance for restaurants offering takeaway and delivery.[2] Similar guides were published by the US FDA, the Greek Veterinary Authority, the Jordanian Food and Drug Agency, the Serbian agricultural inspection, among others. The FAO *Guidance for food businesses* (FAO, 2020[4]) comprised explanations on how to implement basic distancing and provides for cleaning, disinfection and personal hygiene requirements. The International Finance Corporation published a Threats Analysis and Critical Control Points Workbook for businesses to update their food safety management plans (International Finance Corporation, 2020[5]) The FAO Guidelines for livestock production and animal health (FAO, 2020[6]) aims at providing practical recommendations to businesses on how to update good farming practices based on the new situation. Recommendations relate to changes in suppliers, on the need for e-communication with suppliers and buyers, on biosafety and biosecurity measures needed to prevent human contamination with COVID-19 in the farm, on prevention of animal diseases, as well as on cleaning and disinfection and personal hygiene practices.

Control authorities faced various challenges, such as the lack of human resources due to sick leaves and the need to support health systems, reduced testing capacities in laboratories, difficulties to access inspection data by officers working from home, frequent lockdowns' imposed changes in the inspection plans, food safety incidents and high number of complaints regarding foods sold online and increased pressure by customers, media and governments regarding food security and safety (FAO/WHO, 2020[7]).

Since many businesses decided not to reopen, the control plans of food control agencies needed to be updated. The FSA instructed businesses in the UK to notify all reopenings through the notification platform. Only in countries where businesses had the opportunity to notify their opening/reopening online, did control agencies have the time to adjust their control plans. Having a risk-based classification of facilities helped control agencies to prioritize controls in a situation where physical inspections had to be kept to a minimum. As a rule, slaughterhouses did not close and in EU Member States due to *ante-* and *post mortem* control, physical inspections were conducted. Veterinary control authorities continued the inspection of slaughterhouses in Italy and Greece but suspended in other food business operators and performed physical inspection only when food safety incidents occurred. In cases when complaints did not indicate that food safety is jeopardised, control bodies postponed physical inspection and relied on their historical data on compliance of businesses and on proof which businesses provided online.

Risk assessment also helped control authorities to continue monitoring programs for pathogens (animal diseases and zoonosis) as a measure of preventing immediate threats and postponed pesticides monitoring programs [13] Extension services and provision of advice were possible only via phone or the internet. In the UK, FSA did not carry out audits of the voluntary certification scheme Scores on Doors, and instead extended the validity of the already issued marks.

The need for traceability stems from both business and control agencies' side. Should Norwegian salmon producers not have a solid traceability system, they would have not been able to protect their brand from the Chinese control agency´s claim that Sars-CoV-2 virus or its particles were found in one consignment of the Norwegian salmon. **Traceability, once again, proved to be the key for investigating outbreaks and performing efficient recalls when FDA investigated a multistate outbreak of** *Listeria monocytogenes* **infections and linked it to enoki mushrooms imported from Korea. On the other hand, insufficient traceability data, associated with low capacity for strain isolation, prevented Venezuelan control agencies from identifying the source of the** *Salmonella* **outbreak in 500 people.** Furthermore, consumers´ preferences for more organic, locally sourced and sustainable products require detailed traceability data.

Control authorities are faced with the need to control products sold through e-commerce and to perform more efficient recall, both in the conventional and the online supply chains. **The** Canadian inspection agency traced bake food over-fortified with vitamins and sold through the internet to the producer Isagenics and performed an effective recall.[3] In the UK, meat products (lamb, goat, veal, beef), were supplied to retailers and sold directly to consumers by an unregistered and unapproved Wiltshire based vendor through Facebook. The products did not meet the compliance requirements in terms of food hygiene, safety, labelling and traceability requirements and recall was difficult to perform.[4]

Food and supplements sold via the internet were consistently found to be advertised as immune boosters, or allegedly as a prevention means for COVID-19. In many countries, there were no specific regulatory provisions for food sold online, and only regulations dealing with consumers rights regulated this area. FAO clarified the issue and indicated that all food safety requirements automatically apply to food sold online. [15] Due to the change in ingredients and recipes, many businesses, had an issue with meeting labelling requirements. Health-related rules for production of very small quantities are just as burdensome. In Serbia in order to support the operation of newly opened small businesses, regulators issued an exemption from approval for producers of small quantities of food of animal origin, based on the regulation regarding flexible approach to structural requirements. FDA issued temporary policy changes and allowed small farms to sell out of their local community.[5]

Businesses that produce alcoholic beverages required new licenses to start production of medical alcohol, detergents and sanitizers. FAO suggested that instead of going through lengthy approval procedures, in such cases, businesses should be allowed to switch to new production through the process of temporary regulation[13]. This approach was adopted by regulators in the UK, Ireland and by the US FDA to allow the alcohol beverages producer William Grant & Sons to start production of hand sanitizers.

Simplification of food safety regulations and inspection measures: Lessons learned from COVID-19 crisis

This section discusses how to use experiences gained from COVID-induced challenges to improve food safety regulation and the control approach on a long-term basis. Where relevant, this section builds on the *OECD Best Practice Principles for Regulatory Enforcement and Inspections* (OECD, 2014[4]) and its Toolkit (OECD, 2018[5]), both instruments that provide guidance on how to create an effective and resilient regulatory enforcement and control system. They can serve as best practice guidelines to discuss the different lessons learned in the COVID-19 response and as a basis to set recommendations for the future.

Ensure that food safety regulation is adaptable to change

Risk based and flexible regulations: Regulatory systems based on the risk paradigm allowed food businesses to keep their products safe by performing risk assessment of all materials and processes, which needed to change, and by updating their food safety plans to accommodate these changes (Box 4.2). Although such changes were sometimes associated with food incidents, it has to be investigated how prevention of contacts, but also stricter cleaning and disinfection requirements, reflected the frequency of gastrointestinal pathogens transmission through food.

Box 4.2. Creating an agile framework for compositional requirements: an example from Canada

The Canadian Food Inspection Agency (CFIA) has developed an amendment to the *Food and Drug Regulations* to ensure that food compositional standards are more responsive to changes in technology or consumer demand and to ensure that industry innovation is not slowed down. The initiative was developed in collaboration with Health Canada and is part of the Forward Regulatory Plan: 2020 to 2022.

The developed amendment proposes to use incorporation by reference to allow food compositional standards to be maintained and updated in an efficient, timely and transparent manner. This initiative has already been included in the Food Labelling Modernisation consultations from 2013. The initiative

---

is intended to use modern regulatory tools to help foster industry innovation while also protecting consumers from deception and enable more informed purchasing decisions.

A more agile approach to food compositional standards within the *Food and Drug Regulations* is expected to result in a more efficient response from the CFIA to industry and consumer requests for change. In addition, it should contribute to cooperation efforts by facilitating alignment of Canada's compositional standards with international standard setting bodies and major trade partners. The proposal will undergo public consultation in fall 2021.

Source: https://www.inspection.gc.ca/about-cfia/acts-and-regulations/forward-regulatory-plan/2020-to-2022/creating-an-agile-framework-for-compositional-stan/eng/1605050017299/1605050226227.

The crisis revealed that flexibility is not embedded in all regulations. When faced with such a case, regulators must be ready to issue temporary solutions. Systems based on technical regulations, the changes of raw materials, and inputs or technological process can only be amended for a particular product. This is because this is a lengthy and costly process and to solve the problem, regulators must devise new strategies (World Bank Group, 2014[8]). Even the systems based on general principles and risk-based regulations have to use temporary regulation to allow local producers to sell products outside of the designated local boundaries. Examples of temporary regulations introducing easements or additional flexibility include the US FDA's "temporary policy"[6] or extensions of licenses' scope to allow distillers to also produce hydro-alcoholic solution, in a number of countries.[7]

The crisis context has also put in starker light the problems created by fragmentation of the regulatory system, with overlapping layers of rules and institutions. While frequent difficulties are reported by food safety regulators in terms of having sufficient resources to conduct official controls effectively (European Commission, 2020[9]), these resource constraints often reflect institutional fragmentation and duplication, or inefficiency of internal and external processes (e.g. linked to the registration or approval of food business operators, etc.). Such fragmentation also leads to significant issues in terms of regulatory consistency and predictability (Drozd et al., 2018[10]).

Assessing the number of inspectors in charge of a regulatory area (in this case, food safety) is difficult. Indeed, in spite of data on employment in public administrations being generally public, many countries, institutions or services do not keep specific track of inspectors or staff with inspection powers and functions, or do not have consolidated information on all the institutions involved in a given regulatory field. The complexity of regulatory delivery systems where national/federal, state/regional, local/municipal services all can be simultaneously active in a given field makes the task even more challenging. So does the fact that a given regulatory area can be covered by several services, but also that one given service or institution can be, in some countries, active across more than one regulatory field – in which case estimates of resource allocation between these different mandates is not always available.

For these reasons, the OECD Secretariat has so far **been unable to presen**t full data for all OECD members, and even when data is available in some areas, it is not always present for all. The preliminary results of this work show that available resources are often considerable, but may be spread across a number of institutions. They also indicate that there are sharp variations in "intensity" of supervision in terms of number of inspectors by inhabitant, worker, or enterprise, even between neighbouring and otherwise comparable data. This all shows the importance not only of continuing such research and covering more countries and regulatory fields, as well as obtaining more detailed data, but also for countries to conduct such exercises periodically and systematically to review whether the institutional framework and resources are still fit-for-purpose (see Table 4.1).

Table 4.1. Comparison of inspection staff resources in selected countries and regulatory fields

Country	Food Safety	OSH	Env't	Total	Total population	Total businesses	Businesses w/ 10 or more employees	Inspectors/ 100 000 population	Inspectors/ 10 000 businesses	Inspectors/ 10 000 businesses w/ >10 empl.
Austria	2 648	311	120	3 079	8 901 064	410 934	41 940	34.6	74.9	734.1
Finland	810	320	753	1 883	5 525 292	302 901	21 206	34.1	62.2	888.0
France	10 598	2 566	1 890	15 054	67 098 824	3 981 673	160 638	22.4	37.8	937.1
Germany	10 338	5 218	4 374	20 063	83 166 711	2 801 787	361 943	24.0	71.1	550.6
Greece	1 581	629	104	2 314	10 709 739	770 002	29 741	21.6	30.1	778.1
Italy	13 446	6 691	1 002	21 139	60 244 639	3 834 079	176 038	35.1	55.1	1 200.8
Lithuania	720	231	38	989	2 974 090	212 893	13 831	33.3	46.5	715.1

Source: official statistical data compiled by OECD Secretariat.

Overall, the crisis shows that food safety regulations and regulatory delivery should be further developed towards a more risk-based, more flexible and less fragmented system, so as to support businesses and increase their compliance. Regulators need to allow businesses to: a) be aware of, and understand, all key regulations applicable to their business; b) avoid discrepancy in risk assessment and management of the same problem in different regulatory documents; and to c) avoid over-regulation by creating risk based and proportionate norms which will allow fast reaction to crisis while keeping the same level of food safety. This regulatory flexibility follows the OECD recommendation for having norms which are risk-based, more proportionate and simpler in order to allow businesses to address urgent needs.

Rethinking the approach to regulatory delivery of food safety regulation

As a result of lockdowns and other restrictions due to COVID-19, and the reduced capacity in laboratories, inspections and monitoring plans were either partially realized or fully postponed. As per FAO and WHO recommendations (FAO, 2020[6]) (FAO/WHO, 2020[7]), monitoring plans for animal diseases continued while national bodies were left to decide when to resume, and to what extent, activities related to plant protection monitoring. Authorities adopted specific guidance to allow for such flexibility (see Box 4.3).

Engaging third parties to perform inspections and relying on data from efficient food management systems (such as FSMS certification) can help improve resilience of regulatory inspections and achieve overall reduction in the frequency or emergency suspension of checks and visits. In Croatia, the practice of having official veterinarians trained to perform ante and post-mortem inspection in slaughterhouses dates from the 1990s (Miskulin et al., 2012[11]), and, similarly, in Finland[8] veterinarians perform control of animal welfare on a local level. This is in line with Regulation EC 625/2017, which stipulates that official control bodies may delegate audits and inspections (and not actions in case of non-compliance) to a body or a natural person if they have the knowledge, experience, human capacities and equipment, and are impartial and free of conflict of interest when performing their duty according to the instructions provided by the official control body. For the third-party body, additional request is to be accredited. FSMS certificate issued by accredited body is considered as a factor that decreases the risk of a food facility in the Italian province of Lombardy and in Denmark.[9]

Box 4.3. European Commission Implementing Regulation on temporary measures to facilitate official controls on food and feed law in view of COVID-19

In March 2020, the European Commission published an Implementing Regulation on temporary measures given the issues found in the performance of official controls – including the lower capacity and ability to perform physical checks and testing, and to issue and sign official certificates.

Based on the Implementing Regulation, Member States can implement the following measures:

- Official controls may be performed by one or more persons specifically authorized by the competent authority, based on their qualifications, available by any means of communications. Such persons, however, must act impartially, without any conflict of interest.
- Any activity linked to official controls on official certificates and attestations can be carried out on an electronic copy of the original of the document, or on an electronic format of the certificate produced in the Trade Control and Expert System (TRACES). The original of the official certificate must be submitted when technically feasible.
- Analyses, testing or diagnoses may be performed by any laboratory designated by the competent authority on a temporary basis.
- Physical meetings with operators and their staff, combined with official controls and techniques, may be carried out via any available means of remote communication

Source: https://eur-lex.europa.eu/legal-content/en/txt/?uri=celex:32020r0466.

In times of crisis, but also in regular situations when there is shortage of staff and other capacities, prioritisation of control plans should be considered and then preference given to those which prevent and control immediate threats (Box 4.4). Use of third-party audit and inspection results may help realize inspection plans, while at the same time help manage human resources and costs. At the same time, the use of FSMS certificates as an indicator of good compliance will stimulate businesses to implement standards and engage more in self-control.

Box 4.4. Redefining priorities in official controls in the COVID-19 crisis context in Italy

Based on the European Commission's Implementing Regulation of March 2020 (Box 4.3), the Ministry of Health in Italy provided instructions on the control activities by Regions and ASLs (local health authorities) that had to be deferred, and on which others could not be halted – based on their economic impact and the need to guarantee animal well-being.

While controls related to the prevention of African swine fever and avian influenza were initially considered mandatory, scheduled controls for state prophylaxis and activities related to genetic selection plan for sheep and goats were for instance put on hold.

A certain number of controls were maintained, including:

- inspection activities at slaughterhouses
- ante-mortem inspections outside the slaughterhouse in case of emergency slaughter
- official control activities related to the management of the food and feed alert system (RASFF)

In addition to these national measures, Campania provided ASLs with additional guidelines based on the needs and characteristics of the Region. Certain activities were considered non-deferrable:

- follow-up controls for food diseases, export certifications and surveillance of activities related to food safety and veterinary public health

- health checks for brucellosis and tuberculosis (differently from what stated by the Ministry of Health, the Region passed this measure given the status of the emergency in the region)

To reduce risks of contagion, the region of Campania further established that:

- ASLs should adopt measures such as organisational solutions to favor single inspections instead of several control activities, favor more remote checks using national and regional IT systems

- A new procedure for the management of internal audits in the regional health system was to be envisaged, providing for audits to be carried out exclusively remotely with the use of web platforms. This includes "on-site" visits to establishments, farms and facilities being audited in addition to interviews and collection of documents

Source: Ministry of Health of the Italian Republic - Note No. 5086 of 02.03.2020; Note No. 6249 of 12.03.2020; Note No. 10585 of 7.05.2020; Note No. 13173 of 10.6.2020; Note No. 155517 of 10.3.2020; Note No. 163029 of 13.03.2020; Note No. 189403 of 10.04.2020 ; Note No. 198902 of 21.04.2020; Note No. 251127 of 27.5.2020; Executive Decree No. 227 of 01.07.2020.

Ensuring that contingency and crisis management plans remain fit for purpose

A number of authorities responsible for food safety control relied on contingency and emergency management plans to carry out their functions while facing the COVID-19 crisis (Canadian Food Inspection Agency, 2020[6]). Due to shortage of staff, lack of laboratory capacities and other relevant restrictions, those plans had to be changed *ad-hoc* (FAO/WHO, 2020[7]). The COVID crisis was unprecedented and control bodies could not have anticipated the level of disturbances. Experiences show that future emergency control plans should rely more on cooperation between agencies in international trade, to avoid disproportionate measures not based on risk (such as export and import bans).

Future contingency, emergency management and monitoring planning needs to be stress-tested for situations where human and testing capacities are reduced and where logistical problems impair the implementation of any strategy. They have to include the risk assessment of discontinuation of some of the plans and prioritisation of control activities.

Streamlining and simplification of procedures for registration of businesses

The UK experience with online notification of facilities, for instance the re-opening and online change of data in the database of registered/approved facilities, as well as, the effort to simplify measures concerning very small businesses, or businesses trading online, can be seen as a step towards simplification of registration procedures, but also, as a strong support to control bodies´ planning inspections based on accurate data.

Experience e.g. from Ethiopia show that food safety is not jeopardized when licenses for retailers are extended without inspection (FAO, 2020[12]). In Greece, before the Law on licensing was introduced, a pre-condition for starting and operating a food business was the issuance of a license based on the results of an inspection, no matter of the type of business. The license was time limited and required renewal, based on inspection. Simplification of processes to open a food business, and revision of licensing policy proved to boost Greek food start-ups and did not impair food safety.[10]

Food control authorities should engage in digitalisation of food business registration, but also in reducing the number of licenses needed to start a food business (see Box 4.5). Experiences of countries which perform *ex ante* inspections only in facilities which need an approval, should be used by those where *ex ante* inspections require engagement of control staff, unnecessarily extend time for- and increase costs of starting a business.

Box 4.5. Simplifying requirements related to agricultural activities during the COVID-19 emergency: the experience of Trento, Italy

Based on a decree (https://www.politicheagricole.it/flex/cm/pages/ServeBLOB.php/L/IT/IDPagina/15284) of the Ministry of Agricultural, Food, and Forestry Policy in Italy aiming at facilitating agricultural activities as part of the food supply chain the Autonomous Province of Trento issued local regulations with temporary measures to ease the operation of agricultural businesses.

Some measures encompassed the *suspension of legal requirements*, such as the obligation for farmers to notify authorities about works carried out on the land, or the *suspension and extension of deadlines* to comply with obligations, e.g. those related to organic food production with European subsidies.

Other measures implied a *simplification of procedures* to obtain authorisations (e.g. related to discounted agricultural fuel) or an *automatic renewal of existent permits*, such as the authorisation to acquire, use and sell phytosanitary products and to perform consulting activities related to those products.

Another solution was to adopt *alternatives for farmers to comply with the law*, by allowing for instance training to be conducted online. The Province also adopted options for citizens to ensure their own supply of food: citizens not operating as farmers were authorized to grow food for their own consumption without having to comply with the relevant legal requirements.

Improving regulatory delivery has been a priority for the Province of Trento since 2012. These examples of their response to the COVID-19 emergency illustrates this commitment in streamlining and eliminating requirements not needed form a risk-based perspective.

Source: http://www.trentinoagricoltura.it/Trentino-Agricoltura/COVID-19-Disposizioni-ed-informazioni-utili, Decree n. 3318 of 31.03.2020, Vademecum for agriculture activities Covid-19 of 27.04.2020, Provincial law No. 603 of 8.05.2020 and No. 381 of 20.03.2020).

Control should be focused on risks and proportionality and regulation method should be that of responsive regulation

Previous knowledge concerning producers/importers' compliance, may help decide on whether there is a need to physically conduct inspections or if samples for testing will suffice, and, if by doing so, the samples can be prioritised.

Use of self-administered checklists and communicating the results of self-audit to control agencies was recommended by FAO to reduce physical contacts between businesses and inspectors as a direct result of the COVID crisis. The US FDA and the Irish food safety control authorities, for instance, decided to develop such checklists, and use communications from businesses about results of their self-control when discussing with the control agency the reopening of its facilities. In regular situations, control agencies may use self-control data in combination with historical data on business' compliance, to postpone inspection and reduce the frequency of physical control. Physical inspection will remain needed in case of proven food safety incidents.

Risk proportionate control is recommended for border control. Then, trust in producers and importers, derived from good previous inspection records and the level of equivalency between the regulatory and control systems in import and export country, should govern the type and frequency of control (FAO,

2015[13]). The US foreign food facility inspections approach identifies products in compliance with the US regulations and foresees potential food safety problems before products arrive in the USA. The COVID-induced recent Chinese experiences with relaxation of import and export procedures, based on the knowledge of importers and/or producers, is another example, which indicates how import control can better target potential non-conformant consignments.

Information integration improves efficiency of control

Online pre-announcements of shipments are a common practice in USA for all imported consignments. Equally, in EU, the TRACES system allows shipments of animals and food of animal origin to be electronically submitted to relevant inspection bodies. Using electronic certificates or scans and by utilising the interconnectedness between control agencies, inspection clearance processes were accelerated, the number of papers was reduced, and so was the number of contacts between inspectors and owners of consignment. Both practices increased efficiency by reducing costs and allocating resources appropriately the number of persons involved in the control of each consignment was minimised.

The need for healthier food and overall food safety persists. Consumers are expected to seek proof of authenticity (locally produced products, organic, those with controlled and geographical marks), as well as, proof that production has been performed in compliance with environmental sustainability standards. Any claims related to authenticity cannot be investigated if products are not traceable and digital tools can be useful in demand side traceability quests (Baragwanath, 2021[14]). Regulatory agencies should facilitate data sharing (trans-boundary) to accelerate the investigation of potential frauds. More information regarding traceability will ensure better identification of products in the supply chain. The new technology used to capture data, such as Blockchain, can process a huge amount of data and be very useful for tracing such products, but it becomes obsolete when data is inconsistent. The first step towards more consistency will be to support businesses in implementing global standards, such as GS1. The second step would be better data governance. For example, standardized collation of data and sharing of IT platforms for communication of data between control agencies. An example of such a platform is in the EU Administrative Assistance and Cooperation System (AAC system), which helps agencies share data on trans-border non-compliance. AAC system is in line with the Regulation EC 625/2017 which requires collaboration between agencies and information exchange. This system (or some similar platform) may be used to investigate transboundary incidents associated will food and especially with that sold online.

Reality check

Reduction of costs and smart working will be prioritised in the future. Control agencies should explore the regulatory and practical solutions to lay the foundation for smart working. In order to do that, data privacy rules and procedures in agencies must allow inspectors to access the inspections´ data. For instance in Australia,[11] a system of dual identification has been implemented which allows officers to access the inspection databases, via mobile phones. This has been used since 2019 and has been proved to be both secure and efficient.

Stakeholders' satisfaction (businesses and public) should be the object of such reforms. Regulatory and control bodies are expected to safeguard the resiliency of food safety systems, to face all existing and new hazards and non-food related crises likely to affect food security. In the post-COVID era, the tendency to reduce burden incurred by businesses concerning the costs associated with official controls, will become even more prominent given the reduced annual profits. This highlights the need to improve understanding of businesses´self control systems and recognition of process testing results, instead of succumbing to excessive sampling and testing of official samples.

Conclusions

The COVID crisis stressed the need for risk-based, simpler, more efficient, and proportionate regulation. To achieve efficient control, cooperation and data sharing between businesses and control bodies must be prioritised. Since hazards will continue to appear, the capacity of the public and private sector to assess risk and to adapt to new situations by employing the best management solutions, will determine whether provisions can be secured at affordable prices. The new international trade channels require enhanced traceability, but also the use of new technological solutions for data collation and sharing of information.

Online trade requires transparent regulations and collaboration between agencies to monitor food incidents and oversee the management of possible solutions. Increased digitalisation of control operations will help overcome the need for more staff, equipment, support, smart working and accelerated processes, thereby enabling repositories to assess, analyse and estimate data of future trends.

Notes

[1] https://www.food.gov.uk/business-guidance/reopening-checklist-for-food-businesses-during-covid-19.

[2] https://www.gov.uk/guidance/working-safely-during-coronavirus-covid-19/restaurants-offering-takeaway-or-delivery.

[3] https://healthycanadians.gc.ca/recall-alert-rappel-avis/inspection/2020/74287r-eng.php.

[4] https://www.foodsafetynews.com/2021/01/fsa-issues-warning-about-safety-of-meat-sold-on-facebook/.

[5] https://www.fda.gov/media/138316/download.

[6] https://www.fda.gov/media/138316/download.

[7] See e.g. https://www.gray-robinson.com/article/post/2469/fda-and-ttb-temporarily-lift-regulations-governing-hand-sanitizer-in-light-of-covid-19-allowing-distilleries-and-unlicensed-manufacturers-to-produce-alcohol-based-hand-sanitizers and https://www.ft.com/content/e7c02232-67a5-11ea-800d-da70cff6e4d3.

[8] https://www.frontiersin.org/articles/10.3389/fvets.2020.00077/full.

[9]
https://www.foedevarestyrelsen.dk/english/Inspection/Inspection_of_food_establishments/Pages/default.aspx.

[10] Unpublished research by the OECD Secretariat and the World Bank Group.

[11] https://www.foodstandards.gov.au/publications/Documents/fsanz-annual-report-2019-20-accessible.pdf.

References

Baragwanath, T. (2021), "Digital opportunities for demand-side policies to improve consumer health and the sustainability of food systems", *OECD Food, Agriculture and Fisheries Papers*, No. 148, OECD Publishing, Paris, https://dx.doi.org/10.1787/bec87135-en. [14]

Blanc, F. and D. Macrae (2021), "Food Safety Compliance", in *The Cambridge Handbook of Compliance*. [3]

Drozd, M. et al. (2018), *Poland catching-up regions 2 - Safer food, better business in Podkarpackie and Lubelskie*, World Bank Group, http://documents.worldbank.org/curated/en/289891529314549019/Poland-catching-up-regions-2-Safer-food-better-business-in-Podkarpackie-and-Lubelskie. [10]

European Commission (2020), *On the overall operation of official controls performed in Member States (2017-2018) to ensure the application of food and feed law, rules on animal health and welfare, plant health and plant protection products*, https://eur-lex.europa.eu/legal-content/en/txt/pdf/?uri=celex:52020dc0756. [9]

FAO (2020), *COVID-19 and food safety: guidance for food businesses*, https://www.who.int/publications/i/item/covid-19-and-food-safety-guidance-for-food-businesses. [4]

FAO (2020), *Guidelines to mitigate the impact of the COVID-19 pandemic on livestock production and animal health*, FAO, https://doi.org/10.4060/ca9177en. [6]

FAO (2020), *Responding to the impact of the COVID-19 outbreak on food value chains through efficient logistics*, http://www.fao.org/3/cb1292en/cb1292en.pdf. [12]

FAO (2015), *Risk based imported food control Manual*, FAO, http://www.fao.org/documents/card/en/c/caec22a2-b63d-4c27-861d-dd75788ec1d1/. [13]

FAO/WHO (2020), *COVID-19 and Food Safety: Guidance for competent authorities responsible for national food safety control systems*, http://www.fao.org/3/ca8842en/CA8842EN.pdf. [7]

International Finance Corporation (2020), *Food Safety Handbook: A Practical Guide for Building a Robust Food Safety Management System*, Washington, DC: World Bank, http://dx.doi.org/10.1596/978-1-4648-1548-5. [5]

Miskulin, M. et al. (2012), "Food Safety System in Croatia", in *Environmental and Food Safety and Security for South-East Europe and Ukraine, NATO Science for Peace and Security Series C: Environmental Security*, Springer Netherlands, Dordrecht, http://dx.doi.org/10.1007/978-94-007-2953-7_2. [11]

OECD (2020), *OECD Policy Responses to Coronavirus (COVID-19): Removing administrative barriers, improving regulatory delivery*, OECD, Paris, https://read.oecd-ilibrary.org/view/?ref=136_136528-76wdv8q5eb&title=Removing-administrative-barriers-improving-regulatory-delivery (accessed on 15 February 2021). [1]

OECD (2020), *Regulatory quality and COVID-19: Managing the risks and supporting the recovery*, OECD Publishing, Paris, https://dx.doi.org/10.1787/3f752e60-en. [2]

World Bank Group (2014), *Food Safety Toolkit - Risk Assessment, Enforcement and Inspections*, World Bank Group, http://documents1.worldbank.org/curated/en/829181471594886767/pdf/107902-v5-wp-tag-topic-investment-climate-public.pdf.

[8]